ENCOUNTERS!

Encounters! is available at special quantity discounts for bulk purchases for gifts, promotions, premiums and educational uses. For more information, contact: EncountersBookConversation@gmail.com

COVER PHOTO CREDIT: Boris Baran

HARDCOVER EDITION ISBN: 978-1-7362771-0-2

PAPERBACK EDITION ISBN: 978-1-7362771-1-9

E-BOOK EDITION ISBN: 978-1-7362771-2-6

Printed in the U.S.A.

To my extended family, friends and other readers who are atheists or skeptics, who desire deeper peace, purpose and happiness in their lives but feel that something is missing —

May this book bless you and encourage you.

This book is dedicated to you.

CONTENTS

Introduction

Section 1: Stories of Miracles and Supernatural Encounters

Section 2: Evidences of an Intelligent Creator - 143

The Rolex Watch - 147

A. The Intelligent Designs of the Human Body - 149

B. The Intelligent Designs of Animals - 166

PREFACE

I'm glad you are reading this book and hope you'll be as thrilled and encouraged as I am by the following extraordinary stories of supernatural encounters. Almost all of the people in these stories are personal friends of mine. You will notice in their introductions that they come from all walks of life: the CEO of a multinational petroleum company, a member of the Joint Chiefs of Staff, a drug dealer and outlaw biker, scientists, students, doctors, missionaries, business executives, homemakers, military officers, artists, tradesmen, teachers and more.

It has amazed me that so many *encounter* experiences have come from such a small circle of one person's relationships. I can't imagine how many thousands of similar supernatural encounters must occur throughout the world *every day*!

We tend to view life through the limited dimensions of our physical world. Yet, most of us sense that there is more to life than what we can simply touch or see.

Some of our deepest questions and longings cannot be satisfied by things within the physical world: *Why am I here? Is there any true meaning to my life? What is truth? Am I simply the result of a biological act, or is there a specific purpose for my life? When I die, do I simply cease to exist, or is there life after death?*

These are the deep questions that cannot be fully answered outside of a spiritual component. Most Americans say that they believe in a spiritual world, yet why are they surprised when miraculous events occur? Why do we find stories of supernatural encounters, such as those in this book, so mysterious?

I believe it's because most of us live life with little concept of the spiritual realm. Life is hectic and full of daily distractions. Many go through life on cruise control where the days and weeks slip by without any thoughts of the deeper matters of life or spiritual issues. When a supernatural event occurs, it rocks their world.

Some people have an outright disbelief in a spiritual world, so any kind of 'supernatural' event is deeply disturbing. These events challenge their established way of understanding the world and force them to reconsider their worldviews – sometimes with potentially uncomfortable personal implications.

And for those who believe in God, many regard Him as more of an abstract concept than a present and living Being who actively desires to be involved in their lives. Even these believers in God are often surprised when a supernatural event occurs.

I believe there are clear answers to those deeper questions of life above. You *can* know why you're here. You *can* know what your purpose is. And you *can* know what happens to you after you die. I believe the information in this book will bring great clarity and encouragement to you on these topics.

AUDIENCE

I have written this book to encourage two groups of people.

First, to those of you who are atheists or skeptics who sincerely question whether God exists, whether you can know Him, whether there is life after death, whether there really is a heaven and a hell and whether there is any ultimate purpose in your life, my hope is that you will be encouraged, especially in the latter sections of this book.

My encouragement is to read *Encounters* to the end and thoughtfully consider its remarkable contents. This is a profound book. Let the stories of supernatural events open your heart to the fact that there is a spiritual world around you and a Heavenly Father who created you, who loves you, and who desires to reveal Himself to you.

As you read the second, third and fourth sections of the book, which describe overwhelming evidences of an Intelligent Creator seen in science, the human body, animals and nature, I hope you will be further encouraged. I believe this book will bring you great clarity and even answers to many of your questions.

To the second group, to those of you who already believe in God and are seeking to know Him and serve Him better, my hope is that these stories will inspire you to trust Him more deeply and more boldly than ever before. As you consider the numerous examples of your Intelligent Creator seen in the designs of the human body, animals and nature, I hope your faith will be powerfully strengthened as mine has been.

INTRODUCTION

A Personal Struggle

My life has been filled with many blessings, though there have been some difficult trials I and my family have had to walk through. Some years ago, my company was facing a desperate situation involving a bank loan that was about to mature. We had no ability to pay off the loan at that time. In the worst-case scenario, the bank could foreclose on our company assets, which would put us out of business and trigger a series of cascading personal problems.

If the foreclosure scenario occurred, we would lose our home, our savings and probably end up in an apartment with our family of four young daughters. Our children would have to leave their school, plus we could potentially face personal lawsuits from clients whose contracts were unfulfilled. It was a bleak time filled with great uncertainty and fear.

It was during this time, December 2007, that God used my young daughter, Amanda, to remind me of His presence and gave me His indescribable peace, which comes through knowing that He is sovereignly and lovingly in control of my life.

On this morning, I had been praying in my upstairs home office, which overlooks a beautiful small valley and distant mountain peak. In particular, I was reflecting on one of God's names, *El Elyon*, which means *God Most High, Creator and Possessor of the Heavens and Earth*.

At the time, I had been studying several books on the universe, one of which was *The Creator and The Cosmos* by the brilliant astrophysicist Dr. Hugh Ross. In this book, Dr. Ross described how secular scientists, using approximately five different mathematical models, had consistently determined that the universe was *created* an estimated 13-15 billion years ago. According to each of these mathematical models, the universe began in an *instantaneous moment* from a very small size with the explosive release of *unfathomable, pure energy*.

I was reflecting on this thought of instantaneous creation and was trying to grasp how massive our universe has become since its formation and continuing expansion.

For example, just to cross the width of the Milky Way, our own spiral galaxy that contains over *100 billions stars*, it would take *105,000 years* traveling at the speed of light (186,000 miles per *second*).

To put this into context, if you were to launch a space ship at the time Christ was born 2,000 years ago, and were to travel at the speed of light for these entire past 2,000 years (from the height of the Roman Empire throughout the Medieval centuries, through the Age of Enlightenment, the Industrial Revolution and the entire era of modern world history), you would now be *only 2% of the way* across our own galaxy! Try to comprehend this enormous scale.

The closest galaxy to ours is Andromeda. Guess how long it would take to get to Andromeda from earth traveling *at the speed of light*? It would take over *2.5 million years*!

Currently, scientists estimate that our universe contains over *200 billion* galaxies, many of which contain *100 billion or more* stars like our own Milky Way! In fact, some recent discoveries indicate that there are likely over *1 trillion (1,000-billion)* galaxies.

Given that each star, like our sun, is essentially a massive fireball of continuous, roaring thermonuclear explosions, I marveled at the *power* of these stars and the enormous energy and power behind whatever created them in the beginning.

But back to the story of that December morning in my office.

During this moment in the upstairs room, Amanda called. She was eleven years old and was working on her homework downstairs while recovering from a foot surgery. Without knowing of the struggle we were facing in our company or what I was doing upstairs at that moment, Amanda did something she's never done before or since.

She called my cell phone and left a message quoting, then singing, the words to a scripture verse that I had spoken many times from that very room as I would look across our valley to the mountain peak: "I lift my eyes up to the hills. Where does my help come from? My help comes from the *Lord*, the *maker of Heaven and Earth.*" (Psalm 121:1,2)

What a great reminder! *El Elyon*, the One who created the heavens and the earth, the One within whom resides the energy and power of more than *one-hundred-trillion massive fiery stars full of raging nuclear explosions*, was more than able to handle my little problems! The Creator must possess more power than His Creation, and how much comfort this brings when you need His help.

At this point, I began to reflect on the times He has supernaturally revealed himself in my life and in my friends' lives, and I found my faith soaring as I remembered these things. That December morning was the genesis of the book you are about to read.

GETTING STARTED

In the following stories, you will encounter real-life contemporary examples of angels, demons and miracles. In all examples, you will see manifestations of God's power and his love for people.

Why does He intercede in some situations but not others? I honestly don't know. But I do know this: He does intercede *supernaturally* on occasion, and it's thrilling when He does!

It's been my observation that God usually reveals Himself supernaturally when we are in the most desperate of places, where the ability to handle a situation is beyond our control. When our lives are threatened, He can miraculously *rescue*. When our health is in crisis, He can miraculously *heal*. When our relationships are broken, He can miraculously *restore*. And when our finances are desperate, He can miraculously *provide*.

THE SUPERNATURAL REALM

The supernatural realm is all around us, though we can't grasp it in our three-dimensional minds. But just because we can't *see* it or *touch* it doesn't nullify its presence and its reality.

For example, how many times have you heard someone say, "If I can't see it or touch it, I won't believe it." That logic sounds smart, but it won't get you far. Stick a paperclip in a light socket and you'll see what I mean!

It's like the person who stands on the top of a tall building and declares with the greatest sincerity, "I don't believe in gravity because I can't see it," then jumps. He's going to fall only one direction – down! He may have been truly *sincere* in his belief that gravity doesn't exist, but his *sincere belief* didn't affect the *reality*. He was *sincerely* wrong, and the consequences to his life for his sincere – but wrong – beliefs were devastating.

Or take radio waves.

If someone told you that, in the air around you at this very moment were invisible sounds and pictures, would you believe it? Yet, at this very moment as you read these words, you are surrounded by thousands of songs, music, movies, conversations and data streams – all transmitted by radio waves. In fact, they are coursing through your head and body even as you read this. Unless you have the right *receiver*, you can't hear them or see them – but they are there, nonetheless.

So when it comes to the spiritual realm, or to topics like angels and demons, just because we can't grasp them in our physical world doesn't mean that they aren't real and present.

Just saying, "I don't believe in God or in a spiritual world" doesn't affect their realities. That's why the stories in this book are so encouraging to me as they, over and over, reinforce the presence and power of God and demonstrate how He cares for people.

In addition to the stories that comprise the first section of this book, I've added a second section containing amazing facts about the human body, animals and nature that demonstrate overwhelming evidence of *intelligent design* and, therefore, an Intelligent Creator behind the designs. In the final section, I've added provocative thoughts written by my friend Dr. Josh McDowell, entitled "Christianity: Is it Hoax or History."

Josh is a brilliant thinker and former atheist who subsequently became one of the world's leading authorities on the Resurrection of Christ, the accuracy of Biblical documents and historical evidences of Biblical events.

The Resurrection of Christ is the core issue, or lynch-pin, of Christianity: if Jesus actually was resurrected alive from death, something that no other leader of a world religion (Islam, Hinduism, Buddhism, etc.) has ever done, it would underscore the truth of his teachings and and certainly heighten his exclusive claims of deity.

This final section of the book examines a number of issues that atheists and skeptics have raised for centuries regarding whether Jesus actually existed in history, was it possible that he became alive after his death (the Resurrection event), alternative theories on his resurrection, the mental condition of Jesus and whether he was a *liar* or a *lunatic*, the level of accuracy of the Bible's writings and whether they can be trusted, and other topics relevant to anyone who is interested in the search for ultimate truth.

END GOAL

What is the end-goal of this *Encounters* book?

My desire is that you will *encounter* God – that *you* will come to understand, in a deeper way, His amazing love for you and the wonderful plans He has for you and your life.

There's so much He desires to give you!

He offers *hope, purpose, meaning, joy,* and *peace*. He is the only one who can truly satisfy the deepest desires of your heart. He designed you with those deep longings to draw you to Himself, so He can fulfill them as true blessings in your life.

What are the longings that reside in the deepest places of your heart?

They are to *love* and be *loved*... to *know* and be *known*... to be *cherished*... *respected*... *honored*... to have true *purpose* and *meaning*... to have *significance*... to be *secure*... to have in the fullest measures *peace, wholeness* and a sense of *completion*.

Are any of these missing in your life?

Most of us seek to attain these things through human pursuits – money, social status, material possessions, relationships, power and fame. However, *none of these will ever truly satisfy*. If they did, Hollywood would be filled with happy people. But just look at the tabloids!

However, the filling of these deep longings are fully available to you.

Your Heavenly Father offers *all of these deep blessings* to you in abundant measure. And most importantly, He offers *eternal life* to everyone who is willing to accept this most incredible of gifts by trusting and serving Him. Enjoy!

STORIES OF MIRACLES AND SUPERNATURAL ENCOUNTERS

SECTION 1

"Remember the wonders He has done!"
Psalm 105:5

"Lord, I have heard of your fame;
I stand in awe of your deeds!"
Habakkuk 3:2

RAY ANDERSON

WHAT IS SIGOR?

Ray is a carpenter in Florida.
He served many years as a missionary in Kenya.

R ay Anderson is a friend who served as a missionary in Kenya during the 1970s and '80s through Outreach International, an international missions organization. During part of his early time in Kenya, as a young man in his 30's, he was living in Nairobi and was given a vision on multiple occasions of the word *Sigor*. This word would simply appear in his mind, clearly defined and easy to read. He had no idea what *Sigor* meant, whether it was the name of a person, a place, or an object.

Ray scoured phone books for name listings, but found no results. It had no meaning in Swahili or other Kenyan languages. He looked for cities and places with that name, and still no results. For months – and ultimately years – he asked God to give him revelation on what *Sigor* meant, but no answer came.

Then in 1978, about six years after the first *Sigor* visions, the government of Kenya published an updated map of the country. As Ray was looking at it, there, in the western corner of the country, was a city called *Sigor*! With excitement Ray and one of his ministry partners believed this was the answer to their prayers and felt that was where God was leading them to go.

After hours of difficult and rugged travel in their truck, Ray and his partner, Ommani, arrived in the city of *Sigor* just before sundown. Their hopes of finding a motel and restaurant were dashed when they discovered the city was no more than a simple village with dirt roads. No motel and no restaurant. Hungry and discouraged, they felt it would be unsafe to try returning to Nairobi over the rugged road in the dark. They decided they would spend the night in their truck, locked in for safety, and return in the morning.

Ray picks up the story, in his own words.

"Suddenly in the distance of the long shadows of the engulfing dusk we heard someone shouting in English, 'You have arrived, you have arrived, Hallelujah you have come. I knew you would be here!' Ommani and I looked at each other, then at the silhouette of a young Kenyan man running our way. We both wondered, how? Did someone go ahead of us, or send him an announcement of our coming? Impossible! Was he yelling at us? How does he know English? One thing was sure; we were not about to throw cold water on his excitement. We needed a friend, and he was the best thing going. He appeared to know us and was holding out the welcome mat of which we quickly took hold.

'My name is Smuhu, I have been waiting for you for several years now. Look here, I have prepared a place for you to stay while you are in this remote place.' He made a gesture and pointed to a small mud and stick house and motioned us to follow. He scampered around chasing out the goats and chickens and hurriedly dusted off some African made chairs. Through his tears of excitement he said, "Please, you are most welcome in this house." Smuhu called his wife and told her the guests have finally arrived. "Prepare some tea, gather some peanuts and slaughter the chicken, tonight there will be joy in the compound. Our missionary has arrived!' "

Ray was stunned to learn that, three years earlier, the Lord had given Smuhu a vision of a white man who would come to their village to tell them about Jesus. The face of the messenger-man in the vision was none other than Ray Anderson – his funny-looking cowboy hat and all! In the vision, the Lord also instructed Smuhu to build a house and prepare a place to keep the missionary. This small house became affectionately known as, "The Jungle Hilton."

During the following months, God used Ray and Ommani to preach the Gospel to the Pokot people of *Sigor* throughout the dozens of villages in the surrounding area. As a result, thousands of people in this African bush country heard of God's love for them and His plan of salvation, and a great number came to know Him.

RAY ANDERSON

ATTACKED BY A MACHETE-WAIVING AFRICAN

Ray is a carpenter in Florida.
He served many years as a missionary in Kenya.

One day, a man named Joseph came to the mission station for some sugar and clothes. In the course of conversation, Ray became suspicious that Joseph had taken a bicycle loaned to him earlier by the mission and sold it. Later that afternoon, after reflecting on this encounter, Ray and his ministry partner, Ommani, decided to confront Joseph and seek to buy back the bicycle from whomever had bought it.

After an hour's drive into the interior they arrived at Joseph's hut. When confronted with evidence that he had sold the bicycle, Joseph became enraged and ran into his hut. Moments later he came running out with a metal-tipped spear in one hand and a machete in the other!

Shouting at the top of his voice, Joseph screamed, "There is going to be blood shed here today! I am going to kill you, white man, I am going to kill you!" Ommani shouted, "Run Ray, he's going to kill you, he's going to kill you!"

As the native came toward him, Ray began to run around his truck to get away. Despite being a strong and burly man, Ray found himself gripped with fear. Round and around they ran until Ray, exhausted and terrified, tripped and fell by the front right tire of his truck. The crazed African lifted his spear and hurled it! The front passenger door of the truck was opened slightly, and as the spear was thrown, Ray was able to grab the door and pull it open further to use as a shield. The spear punched through the door and partially into Ray's sternum. Ray scrambled to his feet, but he was so exhausted he couldn't run. Now, caught out in the open, the attacker rushed toward him with the machete raised!

Ray relates what happened next.

"My heart was beating in my head as I cried out to God. 'Help me Jesus, help me Jesus!' Something strange began to happen. It was as though someone opened up the top of my head and was pouring 'peace fluid' into me. I sensed as though I was being lifted far above the situation looking down on everything. I stopped and turned around to meet my would-be assassin face to face. Now, with less than eight feet apart, I raised my hands and started walking toward him shouting the name of Jesus! His hands began to quiver. Someone or something was shaking the spear out of his hand, and then the machete. Joseph fixed his eyes above and behind my uplifted hands. Suddenly he let out a blood-curdling scream, then turned away and ran into the bush. I was confused. It was as though he had not even seen me. I turned around and no one was there."

For the next four nights, Ray was unable to sleep. Each night he feared that Joseph, who was familiar with the mission station, would sneak into his bedroom and murder him. Finally, on the fifth day, his attacker did come – but instead of being angry, he meekly and humbly apologized for trying to hurt him! Ray asked, "Why didn't you kill me with your machete?" Joseph responded that there was a huge warrior standing behind Ray with a large drawn sword. At the name of Jesus, God had sent an angelic warrior to protect his child.

But the story doesn't end there. About two years later, Ray was back in Nebraska visiting his home church. Here's the rest of the story...

"While visiting my home church, an older lady came up to me after the service. I knew who she was as she had helped support us through her finances and prayers. Timidly she opened her Bible and showed me where she had written a date. 'Ray, do you know what happened on this date?' My wife, who was with me, said, 'Ray, isn't that the date Joseph tried to kill you?'

As I told her my story, she began to tell me hers. 'Ray, it was during the middle of the night (day time in Kenya); I woke up and sat straight up in bed. I was startled as I could see your face as clear as I am seeing it now, except your expression was one of terror. I yelled out loud, "Oh God, help Ray!" Instantly, I sensed a peace come over me and went back to sleep. The next morning, I wrote this date in my Bible.' When we compared the time of her awakening, it seemed to be the same time I was being attacked by Joseph in Africa.

It's thrilling to realize how important prayer is to God, and that He would awaken one of his saints to pray and participate in spiritual warfare. Even those who are old and frail can be mighty prayer warriors in the Kingdom. And perhaps most thrilling is to realize that our prayers can have *immediate impact* literally around the world!

GENERAL CHARLES KRULAK

A MIRACLE WELL IN THE KUWAITI DESERT SUSTAINS 25,000 U.S. MARINES

Major General Charles Krulak served as Commandant of the U.S. Marine Corps on the Joint Chiefs of Staff and, following retirement, as CEO of MBNA Bank - Europe. He currently resides in Birmingham, Alabama.

I'm honored to share this story of the Miracle Well. During Desert Storm, the United States Marine Corps was ordered to push up the Saudi Arabian coast through the minefields in southern Kuwait and capture Kuwait City. To move 80,000 Marines up that coast, we had to build a logistics support base. We built that base at Kabrit, 30 kilometers south of Kuwait and 30 kilometers in from the Persian Gulf. We picked Kabrit because it was an old airfield that had water wells that provided 100,000 gallons of water a day. The marines needed that much water daily to carry troops into Kuwait.

Fourteen days before the war began, General Norman Schwarzkopf, Commander and Chief of the Central Command, made a daring move called the "Great Left Hook." This sweep of forces flanked the Iraqi army. It was a great move, but it forced the Marine Corps to move the 25,000 marines of the 2nd Division 140 kilometers to the northwest and locate a new logistics base in the Gravel Plains. I was the general in command of this division and its move.

But there was a serious problem. There was no water. For fourteen days we had engineers digging desperately to find water. We went to the Saudi government and asked if they knew of any water in this area, and their answer was no. We brought the exiled Kuwaiti government down to our command post and asked, "Do you know if there's any water in this area?" They said no. We talked to the engineers at Aramco Oil Company who had conducted extensive seismic surveys of the area, and they said there's definitely no water there.

We even went to the Bedouin tribes and the nomads, the people who lived in that area, and said, "Do you know where there's water on the Gravel Plain?" They said, "There's no water there." We kept digging wells hundreds of feet deep to no avail.

Every morning at 7:15 am, during my devotional time, I asked the Lord to help us find water. On the Sunday before we were to enter Kuwait, I was in a chapel Service. We were praying for water when a colonel came to the tent and said, "General, I need to show you something." "Just tell me," I said. "No sir, you need to see this."

We drove down the road we had built through the desert from the Gravel Plains to the border of Kuwait – a road that didn't exist prior to our arrival and over which I had personally traveled probably fifty times. About a mile down that road, the officer said, "Look over there." About twenty yards off the road was a tower that reached fifteen feet into the air. It was a white tower, and at the top of the tower was a cross.

Off the ends of the cross were canvas sleeves used in old train stations to put water into train engines. At the base of that cross was an eight-foot-high pump, newly painted red. Beside that pump was a diesel engine, and beside that, four batteries. On the engine I noticed what seemed to be an "On" button and an "Off" button. I pushed what I believed to be the "On" button, and the engine kicked over immediately.

I called one of my engineers and asked him to test the flow coming out of the pipes. An hour later he said, "Sir, it's putting out 100,000 gallons a day."

What makes this story even more remarkable is that, in our entire division of marines, we had no diesel fuel. Our division ran exclusively on aviation gas and gasoline. Yet next to the pump was a full tank of 1,000 gallons of diesel.

(ADDITIONAL COMMENTARY from the Tampa Bay Times)

Although the water problem was solved, the threat of biological warfare remained as the winds consistently blew from the north to the south, from Iraq into Kuwait, and would carry deadly biological or chemical agents. Some military experts predicted heavy casualties. But Gen. Krulak believes God performed another miracle in answer to prayer.

Fifteen minutes before the 4 am ground attack commenced on February 24th, the wind shifted, blowing from the southwest to northeast. The wind has always blown in the same direction in that part of the world, Krulak said. The shift in the wind neutralized the threat of poison gas because the wind would have blown the gas back toward the Iraqis. Within a few minutes after Schwarzkopf issued cease-fire orders four days later, the wind shifted back to its normal direction. "That is the power of prayer," Krulak said.

As Desert Storm wound down, Krulak held his last prayer meeting of the war at the site of what has been named the "Miracle Well." *(Tampa Bay Times, Oct 11, 2005)*

∽

4

EMMANUEL WOLF

TANK WARFARE MIRACLE IN THE YOM KIPPUR WAR

Emmanuel resides in Birmingham, Alabama

Emmanuel Cohen Wolf grew up in London in a lower-middle class Jewish family. When he was nineteen years old, around 1964, his mother returned from a 10-day trip to Israel. As he remembered, "I had never seen her look so excited and radiant."

"You must see Israel!" she exclaimed.

We learned that a Jewish organization in England would pay the cost for a one-year trip to Israel where we would live in a Kibbutz. So we moved to Israel around 1965. I later became a permanent Israeli citizen.

In October 1973, the Yom Kippur War broke out. A coalition of Arab states led by Egypt and Syria attacked Israel on Yom Kippur. The war began with a surprise joint attack by Egypt and Syria. Both countries' forces crossed the cease-fire lines that had been set in the Six Day War in 1967. Egypt crossed from the Sinai; Syria moved its artillery and tank units onto the Golan Heights, a terribly strategic placement.

The first 24-48 hours were terrifying for Israel. I remember constant rockets coming over and the constant firing of guns. At one point I remember seeing an aerial dogfight overhead between fighter planes.

As an interesting side note, according to a close friend of mine and a personal friend of Israeli prime minister Golda Meir, the first day of the war created tremendous fear and distress in the Israeli government. Late that night, Mrs. Meir called Henry Kissinger. She was crying and pleading with Kissinger for weapons. Kissinger couldn't give her sure support. About 4:00 am that morning, Mrs. Meir was on the verge of taking her life when she received a phone call from President Richard Nixon.

Mr. Nixon agreed to provide all the U.S. military help she'd need, and within 24 hours American support began to arrive. (As historians noted afterward, when Nixon was a small boy, every night before he slept his mother would sit on his bed and read from the Bible. One day she said to him, "Son, one day you are going to do something magnificent for the Jewish people and for their land.")

At the time of the war and this particular incident, I was serving with an artillery unit in the Israeli army. Our unit was positioned in a wadi very close to Mt. Herman near the Golan Heights. We had about 25 artillery pieces. The Syrians were amassed in great force on the Golan Heights, and had they known we were there, we would have been annihilated.

About three miles from our location is where the miracle occurred. This was shared with our entire unit and me by a man who was there.

In effort to try to defend against the massive Syrian armored forces (tanks and artillery) that were gathered on the Golan Heights, a small Israeli tank unit of around 50 tanks raced in the night to get into a defensive position. Their unit was so small that, had they engaged the Syrian tanks, they would have served as little more than a speed bump to the Syrian offensive.

The Israeli tank commander was a non-religious man named David. As they were crossing a wide field, several tanks exploded after hitting land mines.

David immediately ordered a halt as they realized they were in the middle of a Syrian mine field. They couldn't go backward or forward. There were about 500 Syrian tanks several hours away heading their direction, and they must move forward to meet them.

David and his fellow soldiers got down on their hands and knees and tried locating mines with their flashlights and bayonets. It was an impossible job.

About this time, from out of nowhere, a slight wind began to blow from the north. There were no clouds in the clear night sky. Within a few minutes the wind became stronger. In ten minutes, it had grown to such ferocity that the men had to hunker down on their knees with their backs to the wind and hold on to their tanks for protection. Even the tanks were shaking under the powerful force.

After about ten minutes of gale-force winds, the winds began to subside until it was quiet again. When it was over, the wind had blown the surface sands away and revealed all of the land mines! I was only three miles away to the north when this happened and don't recall even experiencing a breeze! (This incident, by the way, was widely reported on by the media after it occurred.)

An interesting thing happened to David a few days later according to the stories I heard. He did meet and engage the Syrian tank force. During the battle he was wounded with shrapnel in his spine. He was evacuated to a hospital for treatment. While on the operating table, the surgeon accidentally cut his spinal cord and paralyzed him.

For weeks, David could only lie on his bed in the hospital. One evening a nurse he hadn't seen before walked into his room or ward and turned on the television. There was a Christian minister on the screen who seemed to look right at David and said, "David, we heard what happened to you. If you can see me, please hear me. Jesus is coming to see you tonight."

The next morning David woke up and began feeling a lot of tingling in his legs. He called his wife to come. The doctor said he was just experiencing phantom pains. David apparently informed the doctor that he was leaving. Within two days he began to walk.

David returned to his home in a Kibbutz and began telling everyone his story about how Jesus had healed him. Neighbors and others got tired of him talking about Jesus and after a period of time asked him to leave. So, he and his family packed up.

On the day he was to leave the Kibbutz, he had no idea where to go and had no money.

His phone rang and it was a man from Seattle, Washington who explained that they had heard about his story. He went on to say that he wanted David and his family to come to Seattle to speak at his university. In addition, he told David that he was sending airline tickets. Several months later David moved his family to Seattle where he later became a preacher.

∼

DON CARMICHAEL

RESCUED FROM ATTACK

Don is a businessman in Alabama

During the summer of 1987, one of my good friends, Tobin, and I traveled the country following college graduation. We spent five weeks of this time in Ft. Collins, Colorado, where Tobin attended Campus Crusade's national Staff Training Conference.

Though not on staff, I volunteered to be part of a Summer Project to help support the conference. While on the project, I became friends with a fellow volunteer named Leighanne, a college senior from Montana. As the weeks passed, we found ourselves spending more and more time together.

One Friday evening, we went up to Horsetooth Reservoir, a beautiful mountain lake that sits in the upper foothills of the Rockies and overlooks the city of Ft. Collins. The spot we chose to sit on and talk was a large boulder about the size of two minivans. We were in a wide-open area next to a gravel road that ran along the ridge overlooking the city. To get to this overlook point, drivers have to exit a paved county road and drive several miles down the gravel road. We were approximately fifty feet off the road and about a hundred feet from my car.

I was wary of being alone in such an isolated place with such a pretty girl. The concern in the back of my mind, especially on a Friday night when people are out drinking, was how well I could protect her if anyone wanted to harm us.

We hadn't seen a car for hours. The only illumination we had as we sat on the boulder was from the stars above and city lights below.

Around 10:30 pm, a car came flying down the gravel road toward us. As the light from the car's high beams picked us up, the driver suddenly slammed on his brakes, causing the car to slide to a stop just past us. My car was in front of this car, but because my doors were locked, there was no way we could reach it safely.

Their car sat idling. Through the windows, we could see two men silhouetted by the city lights behind them. The car looked sketchy and made us nervous.

After perhaps half a minute, both doors opened, and each man got out and looked our direction. They were big, had long hair and looked rough. We could hear them speaking in low tones while their eyes stayed fixed on us (I'm sure on Leighanne more than me). What were they thinking? I'm pretty sure they were planning to overpower me and hurt Leighanne.

I turned to Leighanne and whispered, "Get up."

We stepped down to the ground and stood there, looking at the men as they looked back at us. I quietly pulled a small pistol from my back and chambered a round. We were too far away, and it was too dark, for the men to see what I was doing. Leighanne's eyes got a little big, but being from Montana, she handled it well. I then quietly prayed out loud, "Lord Jesus, please protect us!"

As soon as we finished that short prayer, another car, only the second one we recalled seeing the entire night, crested the hill and began coming toward us. As its headlights began to illuminate the scene, we began to briskly walk toward my car. I kept my pistol palmed, unseen in my hand, and watched the two men; they never took their eyes off us! We jumped in the car and drove away safely.

If that third car hadn't arrived at just that moment, I believe one or more of us would have been seriously hurt or killed.

I'm convinced that second car was divinely directed to the scene at that exact moment for our protection and deliverance from danger.

"Because he loves me, says the Lord, I will rescue him; I will protect him, for he acknowledges my name. He will call upon me, and I will answer him; I will be with him in trouble. I will deliver him and honor him." (Ps 91:14-16)

6

TED MELENDEZ
YOUR EYES HAVE CHANGED!

Ted is a former gang enforcer, drug dealer, outlaw biker and convicted felon. His hard life of anger, violence and crime was radically changed through an encounter with the living Christ, and the outcome has been extraordinary. He lives in San Francisco, California.

B efore a dramatic encounter with Jesus Christ, I had a very rough life. I grew up on the hard streets of Oakland and San Francisco and became involved in drugs, gangs, auto theft and violence at an early age. There was a lot of pain in my life. I was full of anger and would flash into rages. Being short and Puerto Rican, I found plenty of opportunities to prove how tough I was, and I got pleasure from bringing my pain to others.

As a young man, I got heavily into drugs and drug dealing and was part of a notorious violent biker gang. We were into serious stuff. I had martial arts training and became an enforcer for the gang, which gave me plenty of opportunities to use my skills to hurt people. I loved to fight! I showed no mercy. I'm grieved to share that I've hurt many hundreds of people, and many very badly.

During these years before I met Jesus, I was known to cops throughout the rougher parts of San Francisco and surrounding counties. I'm a seven-strike felon. I've served time in San Quentin Prison, Avenal Prison and lots of county jails. I was utterly ruthless and fearless. People would see the darkness of my soul in my eyes and would steer away!

I should have died many times.

I've been in numerous high-speed chases, I've been shot at, stabbed, beaten with baseball bats and jumped by over 20 guys at one point – and though I've come close to death, I've never been seriously hurt. Even in my worst years and behavior, and though I didn't deserve it, God was protecting me and at work in my life.

I ultimately married my girlfriend Tina. I was awful to her – I was violent and dominated her with fear. I wouldn't allow her to have a key to our house and wouldn't allow her to leave without my permission. She finally got fed up and, while I was smoking a joint one morning, walked out for work and didn't come back.

Initially, I was glad to be rid of her. Good riddance!

I went on an all-out party binge. For almost three weeks, I partied around the clock with my buddies in a sleepless haze of drugs and alcohol. But then, I began to realize that I missed Tina and that I truly loved her. I knew I had blown it for good.

I learned that Tina was living with a Hispanic family and was attending their Spanish-speaking church. I didn't know anything about Christianity, but I knew I wanted her back.

A few days later, still partying with hard liquor and drugs, a friend randomly showed up at my house. He'd never been there before. He came in and said, "What's going on Ted? You look awful!"

I explained about Tina, that she'd become a Christian, and I didn't understand what that meant.

He said, "Dude, my mom's a Christian. You can call her and ask her about it."

The conversation with his mom ultimately led me to scheduling a meeting with Pastor Donald Sheley at Church of the Highlands in San Bruno, a city on the peninsula near San Francisco.

During my long first meeting with Pastor Donald, he said, "You've lived a rough life Ted. I bet you wish you could start your life over."

I shared that I'd actually said that very thing many times. He told me that God could give me a new life and wipe my slate clean. Cool! I thought he was talking about God wiping my criminal record clean!

That day in May 2004, in Pastor Donald's office, I surrendered my life to God and committed to serve Jesus. I became a Christian. I didn't feel anything until I went back out to my car.

I saw the big fat joint in my ash tray and didn't even want it. I threw it out, got rid of all my drugs and booze, and was instantly and totally set free. My desire for drugs was completely gone. I took down the dark curtains in my house and replaced them with new colorful curtains and flowers. I set Tina's room back for her.

Unbeknownst to me, God was working in Tina's life too.

That day of my salvation, Friday, she just happened to hear Pastor Donald on a radio commercial and, out of the hundreds of churches in the massive San Francisco Bay area, chose to go to Church of The Highlands in little San Bruno the following night. She needed a church that spoke in English, since she couldn't understand Spanish.

That night, Tina met Pastor Donald, who shared that he had just met me the previous day and that I had become a new Christian.

In the weeks ahead, we began cautiously seeing each other. I begged God to bring her back. I told Him I wouldn't hurt her anymore. She finally said yes, and we began our new lives as new, forgiven, cleaned and beloved people in Christ.

Around this time, I ran into Tina's grandmother who never liked me. She looked at me and said, "Something's different about you. Your eyes have changed."

Shortly afterward, I ran into a man I used to terrorize. He cautiously came forward and said, "You don't look crazy anymore."

Wow. My heart had become transformed! Where did all this crazy love for people come from?

So what's happened since?

I'm now Pastor Ted at Church of The Highlands. I serve our church body as well as reach out to those on the streets and in the rough areas around San Francisco.

Maybe the best example of what Jesus has done in my life can be illustrated by the kindergartners who see me every day at our school that shares a campus with our church – instead of being the feared "Ted The Enforcer," I'm now regularly hugged by these precious children and affectionately called, "Uncle Ted" or "Uncle Teddy Bear!"

DON CARMICHAEL
A SECOND CHANCE - MICHAEL IN OHIO

Don is a businessman in Alabama

Several years ago, I had spent three exhaustive days doing sales training appointments in Tallahassee, Florida with a new rep for my company. We conducted eleven sales presentations in three days in the humid, oppressive heat of summer.

Wednesday afternoon, following our last appointment around 5 pm, I got into my car to begin the six-hour drive back home. Boy was I tired! In less than five minutes, I began to fall asleep on Interstate-10. Desperate for rest or for something that would energize me, I took the nearest exit and pulled into a large truck stop/travel center to get some caffeine.

As I walked toward the entrance to the store, an older man was off to my side, leaning against the trunk of his car, and was trying to get my attention. I knew he needed money, but I ignored him and walked on by. At the counter, I realized my wallet was in my car, so I left my purchase on the counter and walked back out to my car. Again, the man tried to engage me, and again I ignored him. I felt guilty, but I was too tired to be nice. As I passed him the third time going back into the store, I heard him engage the man behind me. "Excuse me sir," he said. "Can you spare some change for a couple of gallons of gas? I'm empty, and I need to get to the next exit..." I remember thinking at that time that it was the first time I'd ever heard that line.

I returned to my car, and as I pulled back onto the interstate, I was overcome with shame and guilt. "God, please forgive me for being so selfish and unwilling to help that man! Please give me another chance!"

That weekend, I flew to Ohio to speak at a golf tournament dinner on Sunday evening. Staying with some friends in the central Ohio town of Dover, Sunday morning found me in a beautiful, quiet and spacious back yard of green grass and flowers.

For an hour or more I spent time in the Word and was moved to re-memorize 1 John 1:9 (*If we confess our sins, he is faithful and just and will forgive us our sins and cleanse us from all unrighteousness*), Romans 8:1,2 (*There is therefore no condemnation for those who are in Christ Jesus, because through Christ Jesus the law of the Spirit of life has set me free from the law of sin and death*), and Galatians 5:1 (*It is for freedom that Christ has set us free; stand firm, then, and do not let yourself be burdened again by a yoke of slavery.*) I felt compelled to diligently memorize those scriptures.

Following the Kickoff Dinner that night at the nearby country club, I was back on the road for a three-and-a-half hour drive to Troy, Ohio for another golf tournament meeting early the next morning.

Around 10:30 that night, I was driving west on Interstate-70 between Columbus and Dayton and noticed that my rental car gas tank was almost empty. I was tired but knew I needed to stop. I passed one exit, but there was no gas station. I was getting really worried about running out of gas.

A little farther, I saw another exit but couldn't see any signs of gas stations. Just I was passing the exit, I saw a gas station sign shining over the tree tops! I whipped the car to the right, jumped the no-man's area between the exit and the highway, and safely made it to the station.

As I was standing at the pump refueling, I noticed a worn-out looking older brown car slowly pull up to the station. The man and woman inside looked just as tired and worn-out as their car, which was dirty, grimy, and filled with trash.

The man got out of the car, saw me, and began walking toward me. "Oh no," I thought. "I'm too tired to get involved..."

"Excuse me, sir" he said. "My wife and I are trying to get to the next exit. We're almost out of gas. Could you spare enough money for several gallons?..."

I couldn't believe it! That line sure sounded familiar – it was the *identical request* I had heard four days earlier in Tallahassee!

With excitement I knew this was a divine appointment, and that God was indeed giving me the second chance I had asked for.

I gave the man a few bills, and he went inside to pay the cashier. As I looked at him through the glass, I had a better idea. I caught his eye and motioned to him to come back.

When he walked up to me, I said, "I'll make you a deal. I'll fill your tank up if you'll let me talk to you about something important."

"Okay," he said, and pulled his car around to the pump.

He told me his name was Michael. I pumped his gas as he leaned against the side of his car.

"Michael, did you know God loves you and has a wonderful plan for your life?" I said. He looked a little shocked, but sincerely interested. I pulled out a small tract that explained how to know God and experience a personal relationship with Christ.

He began sharing the many struggles he and his wife were facing. Then I felt the Holy Spirit give me insight into his situation.

"Michael, I sense you're struggling with drugs and pornography. Are you?" I asked.

He looked completely stunned!

"Yes, how did you know!?"

I then quoted the verses the Holy Spirit had moved me to memorize just that morning!

"Michael, did you know that if you confess your sins, God is faithful and just and will *forgive* your sins and *cleanse* you from *all* unrighteousness? Did you know that He wants to completely forgive you and set you free? He says to you, it is for *freedom* that Christ has set you free; stand firm, then, and do not let yourself be burdened again by a yoke of slavery. Michael, if you ask God to forgive you, His word says that there is *no condemnation* for those who are in Christ Jesus."

Michael began crying.

"Thank you, sir" he said. He admitted he was a Christian but had turned his back on God. He said God must be really trying to get his attention, as I was the second Christian man God brought across his path that weekend.

I prayed for him and his wife, and as I pulled back onto the interstate on that dark, late night, I was filled with exhilaration and great joy. God had indeed given me a second chance and gave me the privilege of being used by Him to touch the life of another person.

～

8

HOWARD HENDRICKS

THE CATTLE ON A THOUSAND HILLS SAVE DALLAS
SEMINARY

*Howard Hendricks served as a beloved speaker and
professor at DTS for over fifty years*

S hortly after Dallas Theological Seminary was founded in
1924, it almost folded. It came to the point of bankruptcy in
1929. All the creditors were ready to foreclose at twelve noon
on a particular day. That morning, the founders of the school gathered in the president's office to pray that God would provide a financial rescue. In that prayer meeting was the renowned Bible teacher
Harry Ironside. When it was his turn to pray, he said in his refresh-
ingly candid way, "Lord we know that the cattle on a thousand hills
are Thine. Please sell some of them and send us the money."

Just about that time, a tall Texan in boots and an open-collared shirt
strolled into the business office. "Howdy!" he said to the secretary. "I
just sold two train-car loads of cattle over in Fort Worth. I've been
trying to make a business deal go through, but it just won't work. I

feel God wants me to give this money to the seminary. I don't know if you need it or not, but here's the check," and he handed it over.

The secretary took the check and, knowing something of the critical nature of the hour, went to the door of the prayer meeting and timidly tapped. Dr. Lewis Sperry Chafer, the founder and president of the school, answered the door and took the check from her hand. When he looked at the amount, it was for $10,000 – the exact sum of the debt. Then he recognized the name on the check as that of the cattleman. Turning to Dr. Ironside, he said, "Harry, God sold the cattle."

(HOWARD HENDRICKS, *Stories for the Heart* compiled by Alice Gray. Portland: Multnomah Press, 1996, p. 272.)

~

9

CHURCH OF THE HIGHLANDS
RANDOM ACT OF KINDNESS PREVENTS SUICIDE

Church of the Highlands is located in Birmingham, Alabama

In the fall of 2007, Church of the Highlands in Birmingham, Alabama, introduced a simple tool to demonstrate God's love to others – with a goal that that demonstration would ultimately point them to God himself. The tool was simple – a small business card that read, "Just a little something extra to show you that God loves you," along with service times and locations on the back.

Though there are several dozen ways this card is used, the most common method is using it in a restaurant drive-thru line. In this setting, when a *Highlands* member pulls up to the window to pay for their order, they also pay for the order of the vehicle behind them and ask the drive-thru attendant to pass along the card.

When first implemented, numerous stories began to be flow back to the church of how this simple tool touched people. In one story, the nearby Starbucks drive-thru attendant said that *eight people in a row* passed along the generosity and paid for the order in the car behind them — and the card was passed along to all eight people! Apparently only one was a *Highlands* member.

But the most sobering and meaningful story was shared by a young woman, not a member of *Highlands*, who was experiencing great emotional distress. Overwhelmed by the struggles she was facing, she had made the decision to commit suicide. She decided to eat her last meal at Schlotsky's Deli on Highway 280 – following that meal she was going to take her life.

As she was driving to Schlotsky's, she cried out to God and asked Him that, if He was really real and if He really loved her, to somehow show her. As she pulled into the restaurant's drive-thru, she was told by the window attendant that the person in front of her had just paid for her meal and had given her a small card. It read, "Just a little something extra to show you that God loves you!"

This simple expression of love touched her so deeply it convinced her that God, indeed, was real and that He knew her and loved her. As a result, she chose life over suicide and emailed *Highlands* to share her story – and to thank some complete stranger for literally saving her life!

∾

10

JEREMIAH CASTILLE

BONE SNAPS BACK AND FORGIVING A MURDERER

Jeremiah was an All-American football cornerback for the University of Alabama and later played in the NFL for the Tampa Bay Buccaneers and Denver Broncos. He played on the last football team coached by Bear Bryant and was one of Coach Bryant's pall bearers. He resides in Birmingham, Alabama.

I was born the eighth of nine children. Our family was poor and grew up in Phenix City, Alabama. I lived in a rough area, worse than any government housing, in a run-down house that had been condemned by the county. I saw a lot of poverty, domestic violence, drugs and alcohol and saw the results. I knew I didn't want my life to go that route. In the second grade, I told my teacher that my dream was to be a professional football player. This was a long shot for a young boy growing up in poverty, but it gave me vision and a sense of direction.

When I was 13, I got suspended for fighting and was crushed when my mom, who was sober that day, said, "Son I'm so disappointed in you." God used her comment to break my heart and make me realize that I needed to change. That summer, at a neighborhood church revival, I heard the Gospel. The most powerful thing I heard was that God loved me!

I became born again that night and experienced immediate radical internal changes. Jesus gave me a new identity of value and worth in Him. His love gave me the power and ability to live through very difficult circumstances as a young man. I began ministering to my mom and dad at that time, and I told my mom, "Mom, one day I'll see you sober!"

None of my siblings graduated from high school; I was able to graduate high school and, through God's grace, receive a football scholarship to the University of Alabama. I became the first person to attend college in my family. While there, Coach Bryant instilled courage, discipline, vision, leadership, character and humility in me in the midst of athletic success and tremendously impacted my life.

In 1983, I was drafted by the Tampa Bay Buccaneers, and the dream I had shared with my second-grade teacher of playing in the NFL had come true. Also, I now had the financial ability to help my mom, so ten years after making my promise, I was able to put her into a rehab facility where she met Jesus. She came out sober and stayed sober for the rest of her life!

In 1988, I was playing for the Denver Broncos. During one practice, John Elway fired a quick bullet pass to me, and the football almost tore off my left index finger. My finger suffered an awful compound break. The bone had protruded through my skin in an "L" shape. A trainer rushed over and looked at it.

"Oh man, you're going to need surgery, and you're going to need a pin in it." He wrapped it in a towel to stop the bleeding.

Our team chaplain had seen it happen and came into the locker room with us. "Hey Jeremiah, can I pray for you?" he asked.

He laid his hand on my toweled hand and asked God to heal my index finger. I felt a powerful heat come over my finger. They rushed me to the hospital and into the emergency room. When the doctor gently pulled away the towel, the bone was perfect! I just needed a few stitches and went right back playing.

One of the most difficult situations happened a few years later on New Year's Day in 1996.

As I walked in the door to visit my mom in Phenix City, she received a phone call and was told that my older brother, Joseph, had just been murdered. I was close to him, and it hit me especially hard!

My mom was deeply hurting, but she said, "The real victim in this, Jeremiah, was not your brother but the person who did this to him." Those words prompted me to pray for the man who killed Joseph.

My mother's response really challenged me – she handled the situation with such spiritual grace and without anger at that person. My heart was moved. I began praying that God would give me the opportunity to share the Gospel with this man.

Two and a half years later, in 1998, a friend involved in prison ministry asked me if I'd go to a chapel in that prison. I went, and when I finished, an inmate came up to me and asked, "Did you have a brother named Joseph?"

I replied yes, and he said, "The guy who killed your brother's in this prison. If you come back next month, I'll have him come."

I experienced surreal emotions... it's actually happening just like I've been praying. I made arrangements to come back, and God led me to speak on forgiveness.

A month later, I'm back in the same room and there... across the room... was the man who murdered my brother. His name was Michael.

I was struggling with anger. I felt God say, "I want you to go to him." I did not want to – I wanted to thrash him! But God was in control...

I walked up to Michael. The Lord said, "I want you to embrace him." I embraced him. I did not want to but obeyed. I felt the Lord say, "Stand right in front of him." I did; man it was uncomfortable. I just stood there looking at him. "I want you to look him in the eye and tell him you forgive him." So I did. "Tell him Jesus loves him." I did.

I finished chapel and left. Four years later, Michael wrote my mom a letter and shared that he had asked God to forgive him for killing her son. Mom actually began corresponding with him and did so until he passed away in prison.

That day in the prison, standing in front of the man who so deeply hurt my family, I encountered the power of God to override my flesh and forgive the unforgiveable. After all, isn't that what God does with us?

∼

11

JOSH MCDOWELL

WOMAN ON THE PIER SETTLES AN IMPOSSIBLE REQUEST

*Josh McDowell is a renowned international author and speaker on Christian apologetics. He has written or co-authored over 150 books in 128 languages. He provides the content in the final section of this book on whether Christianity is a Hoax or History.
Josh and his wife Dottie reside in California.*

Prayer is one area I desire to develop further in my personal life and ministry. It is an area where many people feel inadequate. One of the great needs of our ministry is for men and women to support us in prayer. I would like to share an experience with you that has encouraged me in the area of prayer and perhaps will cause you to be more diligent in this area as well.

I was in California several years ago, shortly after Christmas and the death of my father. I became very introspective as I waited for classes to resume at Talbot Seminary. My mind remained preoccupied after I attended my father's funeral. "I often think how unfair life is sometimes," I remember telling my older brother. "Dad had such a complete transformation of his life that I wish Mom could have shared in it with him." She had died years earlier right after I graduated from high school. I knew it was great that Dad died a Christian. Still, I couldn't help but consider how wonderful it would have been if they could have experienced the joy of knowing Christ together before they died.

My mother had always had a deep respect for church and the Bible. But I wondered why she had never become a Christian believer. No doubt her self-consciousness - for one reason or another - kept her from church. But I didn't know what kept her from trusting Christ. The more I thought about it and worried about it, the more depressed I became. Was she a Christian, or wasn't she? Would I see her in Heaven? I had to know. But how? I thought it was an impossible request, but I prayed, "Lord, you know that I'm miserable.... all I think about is whether Mom died as a believer or not. I have to know, Lord. Somehow, give me an answer so I can get back to normal."

The next morning, I drove to Manhattan Beach on the southern Los Angeles coast, mostly to clear my head and get my mind off my obsession. The weather was pleasant for January, so I parked the car and decided to walk to the end of a pier. Several people were fishing; I all but ignored them. I was too distracted to talk to anyone that day.

I remember staring out into the dark water. An older woman fishing there noticed my brooding and began to talk to me. She was friendly and soon had me engaged in lively conversation. As we talked, we shared where we were from, and to my surprise, she was familiar with my hometown — Union City, MI. But that wasn't all! She said she had a cousin there, in the McDowell family! I almost couldn't believe my ears!

She went on to introduce herself to me as my mother's cousin. She had been brought up with my mother and father in Idaho. She asked me why I was in California, and I told her about attending Talbot Seminary. She was familiar with it and told me she had been a Christian most of her life. She began to recall old memories. My curiosity about my mother piqued, so I asked about her spiritual background. She answered, "Your mom and I were just girls — teenagers — when a tent revival came to town. Your mom and I went every night. It was quite a big thing for our small town. It was so exciting!" Then she added, "I think it was the fourth night — we both went forward to accept Christ." "Thank you, God!" I shouted.

I grabbed her hands and squeezed them. I told her she was an answer to a seemingly impossible prayer! Then I told her about the prayer I had said the day before. We talked on for two hours. Then we exchanged addresses and telephone numbers and promised to visit again.

This was such a specific answer to prayer. I knew God had designed these miraculous circumstances to let me know about my mother's decision. Tears of joy ran down my face as I thought of the reunion my parents were having in heaven. I can honestly say that this episode in my life has caused me to believe in God more and to believe that He wants to answer our prayers — and He will!

∽

12

DR. HELEN ROSEVEARE

MIRACLE DELIVERY IN AFRICA

Dr. Roseveare was an English missionary doctor to Zaire, Africa

One night I had worked hard to help a mother in the labor ward; but in spite of all we could do, she died leaving us with a tiny premature baby and a crying two-year-old daughter. We would have difficulty keeping the baby alive, as we had no incubator (we had no electricity to run an incubator) and no special feeding facilities.

Although we lived on the equator, nights were often chilly with treacherous drafts. One student midwife went for the box we had for such babies and the cotton wool the baby would be wrapped in. Another went to stoke up the fire and fill a hot water bottle. She came back shortly in distress to tell me that in filling the bottle, it had burst. Rubber perishes easily in tropical climates. "And it is our last hot water bottle!" she exclaimed.

As in the West, it is no good crying over spilled milk, so in Central Africa it might be considered no good crying over burst water bottles. They do not grow on trees, and there are no drugstores down forest pathways. "All right," I said, "Put the baby as near the fire as you safely can, sleep between the baby and the door to keep it free from drafts. Your job is to keep the baby warm."

The following noon, as I did most days, I went to have prayers with any of the orphanage children who chose to gather with me.

I gave the youngsters various suggestions of things to pray about and told them about the tiny baby. I explained our problem about keeping the baby warm enough, mentioning the hot water bottle. The baby could so easily die if it got chills. I also told them of the two-year-old sister, crying because her mother had died.

During the prayer time, one ten-year-old girl, Ruth, prayed with the usual blunt conciseness of our African children. "Please, God," she prayed, "send us a water bottle. It'll be no good tomorrow, God, as the baby'll be dead, so please send it this afternoon."

While I gasped inwardly at the audacity of the prayer, she added by way of corollary, "And while You are about it, would You please send a doll for the little girl, so she'll know You really love her."

As often with children's prayers, I was put on the spot. I honestly said, "amen" at the end of the prayer, but I just did not believe that God could do this. Oh, yes, I know that He can do everything. The Bible says so. But there are limits, aren't there?

The only way God could answer this particular prayer would be by sending me a parcel from the homeland. I had been in Africa for almost four years at that time, and I had never, ever received a parcel from home. Anyway, if anyone did send me a parcel, who would put in a hot water bottle? I lived on the equator!

Halfway through the afternoon, while I was teaching in the nurses' training school, a message was sent that there was a car at my front door. By the time I reached home, the car had gone, but there, on the verandah, was a large 22-pound parcel!

I felt tears pricking my eyes. I could not open the parcel alone, so I sent for the orphanage children. Together we pulled off the string, carefully undoing each knot. We folded the paper, taking care not to tear it unduly. Excitement was mounting. Some thirty or forty pairs of eyes were focused on the large cardboard box.

From the top, I lifted out brightly colored, knitted jerseys. Eyes sparkled as I gave them out. Then there were the knitted bandages for the leprosy patients, and the children looked a little bored. Then, as I put my hand in again, I felt the... could it really be? I grasped it and pulled it out – yes! A brand-new, rubber hot water bottle!

I cried.

I had not asked God to send it; I had not truly believed that He could.

Ruth was in the front row of the children. She rushed forward, crying out, "If God has sent the bottle, He must have sent the doll, too!" Rummaging down to the bottom of the box, she pulled out the small, beautifully dressed doll. Her eyes shone! She had never doubted!

Looking up at me, she asked, "Can I go over with you, Mummy, and give this doll to that little girl, so she'll know that Jesus really loves her?"

That parcel had been on the way for five whole months!

It had been packed up by my former Sunday school class in London, whose leader had heard and obeyed God's prompting to send a hot water bottle, even to the equator.

And one of the girls had put in a dolly for an African child – five months before – in answer to the believing prayer of a ten-year-old girl to bring it "that afternoon."

"Before they call, I will answer!" Isaiah 65:24

～

13

FRANK BARKER
FROM NAVY FIGHTER PILOT TO PASTOR

Rev. Frank Barker is a former U.S. Navy fighter pilot and the
founding pastor of Briarwood Presbyterian Church.
He resides in Birmingham, Alabama.

In 1954, before coming to know Christ, I was an officer in the U.S. Navy going through flight training school at Barin Field in Foley, Alabama. As a young fighter pilot, I had heard the Gospel but had been unwilling to commit my life to Christ. I was full of vigor and had too many wild oats to sow.

I returned to my home in Birmingham one weekend to spend time with some buddies. We had a wild weekend of partying and sinning. Sunday night, as I was driving back to the base in Foley, I struggled to stay awake. I downed Coca Colas, one after another, until I was about to pop. Despite the caffeine, I ultimately fell asleep at the wheel.

When I dozed off, the highway turned left – and I went straight! The car left the pavement and, fortunately, hit a dirt road. I jolted awake as my car spun out of control in the dirt. I found myself frantically trying to avoid a crash as the car skidded back and forth. Finally, I gained control of the car and brought it to a stop.

There, in the sudden stillness of the night as the dust began settling, I noticed that directly before me was a large sign. By the illumination of my headlights, I read the following words from the Bible emblazoned across the sign: "For the wages of sin is death!"

I had been sinning all weekend. As I looked at that sign, the Lord got my attention and, from that point forward, I began to get more serious about my relationship with Him.

Several more years passed until I truly committed my life to Christ. During this time, I had other close calls, especially related to flying off and on aircraft carriers. When these close calls would happen, I'd briefly get serious about my faith again.

I had been praying and asking God to make me different. One time, I felt He was really listening and felt He was saying, "Do you really want to be different, or are you just talking?" I remember feeling that, if I said "Yes," God would do something from the inside to truly change me. I realized I was resisting God. Didn't He love me? Why would I resist Him?

At this point of decision, I realized that if I went forward in a relationship of serving God, my entire lifestyle would change. The peer pressure would be intense. I was planning a wild party weekend with a lot of my Navy pilot buddies. I told God that, if He got me out of this quandary, I'd follow Him. He did.

Six months later, I felt God was calling me to attend seminary. I was preparing to go on a nine-month cruise to the Pacific then leave the Navy. I took a two-week vacation prior to the cruise and drove from Miramar Naval Air Station in San Diego (where I was currently stationed; it later became the Navy's Top Gun fighter school) to Birmingham. I drove to South Highland Presbyterian Church to find out more about which seminary to attend.

Each morning, I would drive to South Highland Church to meet my childhood pastor, Dr. Frank Mathes. But I couldn't bring myself to go in the church! Morning after morning, I'd wait in the parking lot, then ultimately chicken out and drive off. I remember thinking how I'd be a lot more comfortable talking to Dr. Mathes on a golf course than in the formal setting of his church.

A few days later, I was playing golf at Birmingham Country Club. I sliced into a tree on the side of the fairway. As I went into the trees to find my ball, there he was, bending down to get his ball as he played from the opposite fairway. We finally had our meeting!

When my Pacific cruise ended nine months later, I returned to the states and went to seminary. I still wasn't a true Christian. I'd been in seminary about a month when, one day, my roommate walked in the room. He was graduating and asked me if I'd take over his position as pastor of a small church in Oxford, Alabama. So, I became pastor of Dodson Memorial Presbyterian Church while in seminary.

About a year later, still in seminary, I was deeply reflecting on *exactly what is a Christian*? I'd seen the billboard sign saying, "The wages of sin is death." As I reflected on that passage in Romans 6:23, I considered the rest of the verse: "...but the *gift* of God is eternal life in Christ Jesus our Lord." Eternal life, salvation, was a *gift*!

I studied Ephesians 2:8, which says, "For it is by grace you have been saved, through faith – and this not from yourselves, it is the *gift* of God – not by works, so that no one can boast." I realized salvation was a *gift* from God – beyond anything I could do to earn it.

So finally, as a seminary student who had come to know a lot *about* God, I finally came to *know* God. By surrendering my life to Him and His lordship over me, I finally became a true, born-again Christian. And my love for other people soared.

I'd have to say that the past 40-plus years since making the decision to follow Christ has been far more exciting, rewarding, and fulfilling than anything I ever imagined. I encourage you to make that same decision and receive the free gift of salvation God is offering you.

～

14

BEATTY CARMICHAEL

MIRACLE HEALINGS OF RANDOM PEOPLE

Beatty is a businessman in Birmingham, Alabama

My passion is to help bring people into a life-changing relationship with God. One of my greatest joys has been to discover how powerfully He still works through miracle healings that demonstrate His presence and love for people.

Several years ago, I began to regularly go to nearby Walmart stores on Saturday mornings to pray for people with physical ailments. God desires to heal, and He uses divine healings, in part, to reveal Himself to the person who is suffering.

The following are some stories from the literally *hundreds* of people over whom I've had the privilege to pray with healings occurring. I used to keep a detailed journal description of each healing I observed but stopped some time ago after surpassing more than 200 accounts.

. . .

• I WAS ATTENDING a local Bible study led by Frank Barker, a long-time family friend and retired pastor of my church. Most of those in the study were elderly people. I shared my testimony of how God often moved in my prayers of healing over people, and a lady named Lou came up wanting prayers for her eyes. Lou had macular degeneration and could barely see.

We were standing by a table with people seated at it, and she could not see any of the faces clearly. About forty feet away was a fireplace containing a mantle with decorative accents on it, and everything was completely fuzzy to her. So, I prayed for her, and each time I prayed, her vision became progressively sharper. She began to see faces clearer. We kept praying, and she started to see the accents on the mantle more and more crisply. After about ten minutes, we stopped praying and she said that her vision was the clearest it had been, "in a long, long time."

• I WAS in Walmart looking for people to pray over when I saw a man named Greg. He was in the children's clothing area with his wife. It looked like he had a bit of a limp, so I asked him if he had any pain and if I could pray for him about anything. He said he had back pain and showed me a long scar on his back. He had undergone seven surgeries on his back, and the doctors said there was nothing they could do for the pain.

I asked him what pain level it was, he said that it was typically not too bad, but right then it was an 8 or 9 (on a scale of 10). I prayed for him and asked him to check it out, and he had this bewildered look on his face like he couldn't believe it – all the pain had completely disappeared. I then asked him if he knew Jesus and he said yes, he's been a Christian for years. I asked if he had any other pain. He said his left knee was bothering him. It was currently a 6 or 7 on the scale. So I prayed for his knee and all that pain disappeared, too!

. . .

• BEA WAS A LARGER woman in bright blue sweat pants riding a scooter at Walmart. Her son David was walking behind her with a walker. She was racked with pain. She had broken her hips, and an MRI scan showed that her hips were fractured and that one ball joint was jammed up into her hip socket. Her pain level was an 8 on the scale (on a scale of 10). I prayed for her one time, and all the pain completely disappeared. She stood up out of the scooter and moved around and could not feel any pain. She started swooshing her rear-end back-and-forth to the point that her son said she needed to stop or people would think she's doing something funny.

She also had neuropathy in both hands. She wore gloves because the pain was so bad. The pain level was a 10, and it stayed at 10. She would never shake someone's hand because it hurt so bad. So, I prayed for her hands, and all of the pain left. All of the sensitivity went away, except for a little bit of feeling like tingling granules of sand. She was surprised that she would let me hold her hand because no one had held her hand in such a long time.

∼

15

CHRISTIAN ATZ
THE LADY IN THE PURPLE HAT

Christian is a high school student] in Birmingham, Alabama.

One Saturday morning several years ago, Beatty Carmichael met with some friends in preparation for a day of praying for people at a nearby Walmart. In this small group were his friend Aaron Atz and Aaron's ten-year-old son, Christian. As they prepared to pray, Christian shared that he had a very painful bone protrusion in his ankle that sometimes hurt so badly he would have to stay home from school. He asked them to pray for him, and as they did, Christian saw the protrusion retract with a popping sound and felt immediate relief!

As the group continued praying, Beatty encouraged them to ask the Lord to provide visual guidance on specific people He wanted them to pray over at Walmart for healing. Christian said that he saw in his mind's eye a vision of a "black lady riding a blue scooter." Beatty shared that this description could fit way too many people at

Walmart, and could he be more specific. Christian mentioned that she was wearing a purple cap.

With another request from Beatty to be more specific, Christian began to provide extraordinary clarity. He described that she wore a knitted cap that looked homemade and had red, purple and blue threading, like a beanie cap. He then described in detail the jacket she was wearing – red with white stripes running down both sleeves and had a hood on the back – plus a few other unique identifying features of her attire.

When the group arrived at the Walmart parking lot, one of the members said, "Look, there she is!" They saw a middle-aged African American woman riding a blue scooter wearing a purple homemade knit cap, a red jacket with a hood and white stripes down each sleeve just as Christian had described in his vision!

16

TROY CARMICHAEL

NIGHT ATTACKS

Troy is a businessman who resides in Alabama

Troy Carmichael was a student at Auburn University and was part of a student ministry called Campus Vision. Campus Vision sought to bring together the numerous Christian student organizations to build unity and help stimulate a greater overall campus ministry impact.

During one fall quarter, Campus Vision hosted a large three-day campus-wide event called *Celebration*. Two weeks prior to the event, Troy and a group of fellow students formed a 24-hour prayer chain to be in continuous daily prayer until the *Celebration* event was concluded. Troy's slot was from 2:30 am – 3:00 am.

During this two-week period, Troy, who lived alone in an apartment off campus, awakened around 2:30 am with an extremely dark and oppressive evil presence in his room. His mind was reeling with thoughts of death and suicide – and as he awakened further, he saw a white vaporous form by his bed! Knowing it to be demonic, Troy rebuked it in the name of Jesus and commanded it to leave. After several minutes of intense prayer and quoting scripture, the demonic presence left.

The following day, Troy shared this experience with his discipleship group leader, Scott, and his pastor from a nearby Baptist church. They met at Troy's apartment and prayed over each room. They then anointed his bedroom doorway with oil.

Later that evening during his prayer slot, Troy was shaken by a sudden pounding on his bedroom door and sensed it was a demon!

He sat up in bed and rebuked the spirit, feeling the flow of God's spirit course through him, and commanded it to leave.

The pounding ceased immediately – and no demonic presence entered his room.

Troy got up, opened the door, and went into the living room to check the locks on his front door. They were all securely locked from within... and no one was in the apartment. The harassing spirit was able to reach his door but couldn't cross the threshold that had been anointed with oil!

～

17

VICKI CARMICHAEL

THE INDIAN WOMAN'S VISION

Vicki is an artist and homemaker in Birmingham, Alabama

One evening in 1995, my husband Don and I were home watching TV and came across a sermon being preached by a respected pastor in Los Angeles named E.V. Hill. Dr. Hill was sharing how he had recently lost his wife to cancer and was explaining, with great emotion, how difficult his journey of grief had been in the aftermath.

He used the analogy of how it felt like being in the midst of a deep, dark night that seemed to last forever. During the depth and darkness of the night, when you think the night will never end, the dawn always comes. It may seem like forever, but it always comes.

"You've got to keep walking, even when you feel like the pain will never go away!" he proclaimed. His point was that, when you feel like your struggle will never end, God always shows up and will improve your circumstances.

My husband and I looked at each other. We both felt it – a clear, strong sense that this message was directly for us and was a warning that we were about to walk through a season of struggle. It was such a direct and foreboding feeling! Don had recently discovered that his business partner was siphoning money, engaging in unethical business practices and putting our company in jeopardy. He was in the midst of dealing with these problems when the sermon occurred.

Over the next six weeks, we both continued to receive scriptures and other signs that seemed to indicate we were approaching a difficult season. Something was about to happen.

At about the six-week mark from the E.V. Hill sermon, Don and the board of directors had showdown meetings with the partner. On this particular day, the partner, who was the majority owner of the company, fired my husband, our company accountant and our board. Though we didn't know how challenging it would become, our forewarned season of trial had begun. It would ultimately last three-and-a-half years.

During this time, I was involved in a weekly local Bible Study Fellowship gathering of about a hundred ladies from all different church backgrounds across our metro area.

Several weeks following the showdown, as I was trying to adjust to our current unemployed circumstances, a new woman attended our meeting and joined my small sub-group of about ten ladies. She was Indian, and when she saw me sitting across the table, she became very excited. In her Indian accent she exclaimed, "Oh my goodness, you're the one! You're the woman I've been seeing in my dreams!"

I asked her what she meant, and she described how she had been seeing me in a recurring dream that she had been having over the past several weeks. She described seeing me lying on a bed, then sitting up with expressions that told her I was full of worry and anxiety. Because of the dream, she told me that she had been praying for me for these past two weeks.

Out of curiosity, I asked her what I was wearing in the dream, and she said, "You're dressed in a beautiful white wedding dress."

Uncertain what this meant, I asked her if she understood the meaning.

"Yes," she said. "It means you are clothed in God's righteousness as the Bride of Christ."

I knew that this term, "the Bride of Christ," referred to all Christian believers, but I also received this as a personal term of endearment.

This encounter encouraged me so much and has lifted my spirits for so many years! I didn't need to worry or be fearful; God Himself was reminding me that I was His, that I was His bride, and therefore I knew I could trust Him for all our needs.

And what happened to the Indian woman?

We had never seen her before, and we never saw her again.

Whether she was an angel or simply a person, either way, I believe she was sent for the specific purpose of encouraging me that God was at work, and that I didn't need to be fearful with worries or anxieties.

18

DR. TOM DOOLEY
PROPHETIC VISIONS ABOUT 9/11

Tom is a biotech executive with years of experience as a biomedical research scientist. He has written three books and founded two biotech companies. He resides in Birmingham, Alabama.

The tragic events of the 9/11 Islamic terror attacks against our country have caused many people to question, "How could God permit such a thing to happen?" Many pondered whether anyone had any kind of advance warning. The surprising answer is yes.

I don't know how many people around the world experienced these events from a perspective similar to mine... While prostrate on the floor in prayer several weeks prior to 9/11, I definitely knew in my spirit that something very grave was about to happen. I was drawn into an intense two weeks of prayer leading up to 9/11. I could barely get any work accomplished. I couldn't concentrate, as my thoughts turned exclusively to intercession. But I wasn't sure what it was that needed prayer.

On the weekend prior to 9/11, I had a phone conversation with my dear business executive/prophetic friend, John Manwell, in Liverpool, England, and mentioned that something very serious was afoot. I asked for his impression.

John shared a remarkable vision that he had seen about skyscrapers falling from the sky, an accompanying earthquake and a major stock market collapse. I asked him where it was going to happen, and he believed it was likely in London, where we both had formerly lived. I was in agreement with him in my spirit overall, but I questioned whether the city was, in fact, London. Although John felt it was, I did not sense that London was in any imminent threat.

Then, a few days later on the morning of September 11th but before the towers were hit, I was sitting in my chair at home reading my Bible before going to work. Without any prior notice, I experienced a brief instantaneous trance. The room disappeared, and I clearly saw a large black flying winged object, like a large bird (or perhaps a plane) in form. It was flying directly toward me at high velocity and was about to crash. In shock I shouted out loudly, "No!" and the trance disappeared.

It lasted only several seconds. I sat there with my heart racing in my chest, not knowing exactly what this flying black bird message meant. However, I did know for sure that it involved the *angel of death* and that human lives were at risk. I began to meditate on the term *angel of death*. People were soon to die.

Without knowing it, that was the precise time when the four commercial jet planes were beginning to be hijacked by suicidal Islamists in the eastern USA. Within a few short hours that morning, tragedy and chaos would hit the East Coast. The ground would tremble upon impact of the collapsing buildings and a stock market collapse would quickly follow... just as John Manwell and I, together, had foreseen in advance.

〜

19

DON CARMICHAEL

PROPHETIC DREAM ON A CRUISE SHIP REVEALS A CRIMINAL
ASSAULT

Don Carmichael is a businessman in Alabama

On Valentine's Day several years ago, my wife, Vicki, and I were on a large cruise ship on the Caribbean Sea between Grand Cayman Island and Cozumel, Mexico. That night as we sailed across the sea, I had an unusual and vivid dream.

Over breakfast the next morning, I shared the details of the dream with Vicki and how a central character in it was one of our daughter's dear friends named Makenzie. Makenzie is like an adopted daughter and extension of our family.

"That sounds like more than a normal dream," said Vicki. "I wonder if it's prophetic? Why don't you email a description of it to Mak and see if anything resonates with her?"

It was a great idea. I emailed her the following description from the ship and waited for her response. Here was the dream's description:

Makenzie and I were at our family country home along with a group of family and friends. She and I took a short walk down the driveway next to the house by the rope swing tree toward the road. It was night-time, and the sky was filled with stars.

As we walked, I saw three lights high in the sky moving among the stars in our direction at high speed. They were the same size as the other stars. At first I thought they were satellites, but then saw five additional lights, behind the first three, flying in a tight V-formation. I then had immediate knowledge that they were U.S. fighter jets and heard the deep rumble of jet engines as they passed.

Out of one of them a small clear plastic ball/orb came down and landed just inside the woods below the rope swing and just on the other side of a wire fence. It emitted a strobing blue and red flashing light. The ball was clear and about the size of a soccer ball.

As we looked toward the ball, a young girl appeared in a full-length black dress. She didn't speak and appeared to be around five years old. She silently moved around inside the woods by the flashing ball. I asked Makenzie if she knew who that girl was, and she said, "Yes, that's me as a little girl."

I then had a word of knowledge in the dream and knew that Makenzie had been sexually assaulted as a little girl, probably in the woods at night time. I clearly sensed that the little girl represented Makenzie in the past, but that she was no longer that little girl.

A clear inaudible voice, which I knew to be God, said, "Tell her that she is completely healed, and that I am restoring her." At this point, the scene shifted to inside the house, and the details faded away.

To my surprise, a few hours later, I received an email back from Makenzie saying that the dream was definitely for her!

She explained that she and my daughter had been spending time together several months prior, and that my daughter had made the comment that Makenzie was exhibiting some characteristics of a person who had been sexually abused.

Makenzie said she had no knowledge of these abuses but began to earnestly seek God for understanding and healing. Over the previous several months, He had begun revealing memories of some of the abuses that had happened.

Makenzie then began to give insights about the dream. Her father was an Air Force colonel and flew fighter jets. He was abusive to Makenzie and her mom. She had learned from her mother that her dad would often get drunk and violent.

Here's the interpretation:

In the dream, our family home represented a loving, safe and healthy home. The fighter jets in the sky pointed to her father. The little girl just inside the tree line was her as a young child. The full-length black dress indicated grief, shame and the mourning of innocence. The pulsating red and blue lights in the clear plastic orb represented police lights at a crime scene.

Makenzie believes she was likely assaulted around five years old in the woods just behind their base housing in Korea. She went on to share that while in Korea, her dad had been sent on an assignment out of the country. Her mom seized that opportunity to grab the children and escape back to the U.S. to a secret location, far away from him and his abusiveness.

Through God's grace, He gave me (a healthy father figure in her life) a dream to show Makenzie that He heard her prayers and communicated the clear message to her that, "You are completely healed, and I am restoring you." Wow.

20

TIM TAYLOR
GOD ORCHESTRATES AN AMAZING LOVE STORY

Tim is a businessman in Kokomo, Indiana

I knew I was going to marry Susan two weeks after our first date. I'm not sure that I believe in love at first sight, but I do believe that God had planned for us to be together. I remember being smitten the first time I saw her.

Before I had arrived at college, I had determined that I would remain true to my faith in God. I was on a mission to prove that I could keep my values and not give in to the temptations that permeate the freedom of college life.

I joined a fraternity my freshman year. I was a freshman "pledge" and had to drive a carload of older fraternity brothers to Florida at Spring break. I didn't drink, so I was the designated driver. I didn't party with the guys while I was down there and chose to hang out at the hotel. I ended up meeting a sweet older couple who were staying in the hotel room next to ours. Every day I made it a point to apologize for my loud obnoxious roommates.

The Stewarts were very forgiving, kind and interesting. We ended up spending a lot of time talking together. As only God would have it, they were from my hometown in Kokomo, Indiana, and we knew many of the same people. They even had a granddaughter that they wanted to fix up with me:

"Really, how old is she?" I asked.

"Oh she's a freshman," Mrs. Stewart replied, "and you two would be perfect for each other."

By this time, I was amused, and mildly interested that this cool grandma wanted to fix me up with her granddaughter.

"What school does she go to?" I asked, wondering if her college might be near mine.

"Kokomo High School. She's 14," Mrs. Stewart replied.

Ugh! "Kokomo High School!" I retorted laughingly. I certainly wasn't interested in a girl just barely out of junior high.

"Well, I'm just going to be praying about it anyway," said Mrs. Stewart. "You two should be together!"

I gave her a hug and thanked her for the compliment. A bond was created that week, and I really ended up having a great Spring Break with two old folks. They took me to dinner several times, and Mr. Stewart even let me dance with Mrs. Stewart one night at the club where we dined. It's not the normal ideal Spring Break for a college freshman, but overall, I had fun!

The more I learned about God, the more He opened doors for me at school. By my last year, I had become known as the "chaplain" of the Fraternity. Guys would come to me often and ask for prayer about things in their lives. One particular freshman pledge came to me very disturbed one night about a situation he had gotten into. He had talked his high school girlfriend into having sex with him and she became pregnant. He wasn't sure he loved her and the thought of having a baby was terrifying. Although He didn't have a relationship with God, he came to me for prayer and advice.

He felt guilty about manipulating his girlfriend, but he certainly wasn't ready to be a dad. As God would have it, (coincidently?), he was from my hometown and had known about me most of his life. He grew up not far from where I did, and he looked up to me. I think he might have imagined that I would give him some kind of permission to just move on with his life. Instead, I told him to go back, marry this girl, become a good husband and father to his child. I told him about the freedom God gives us to choose, and that many times it creates painful situations. I had no reason to believe that he even knew God, but if wanted me to pray, I was going to pray that he and she would come to know God through this experience.

I really felt sad for this guy, but I remember most of all being drawn to his girlfriend. He showed me a picture of her, and I visualized the enormity of his decision on her life and that of her baby. They had been deceived into living for the moment satisfying their own desires. Now it produced a life-changing responsibility for them both. He told me that he wanted her to have an abortion. I was thankful that God had put me in this situation to be there for this guy, and they ultimately decided against an abortion.

I was unaware at the time, but God had been weaving His plan for Susan and me way before we met.

Five years later I was living back in Kokomo and, through a random encounter involving playing tennis with a friend who suggested adding his younger sister as a fourth doubles partner, met a girl named Susan. We immediately connected, and several months later she told me she wanted me to meet the most spiritually significant person in her life – her grandmother. We went to her grandparent's house where I met her grandparents for the first time.

When I saw her grandmother, she looked strangely familiar.

With a big smile, she said rather matter-of-factly, "I told you that I would be praying for you two to be together."

Wow! I was dumbfounded!

I realized then that I had not only met her grandmother before, I had also danced with this special woman five years earlier on a Spring break trip to Florida. God is Amazing!

I also found out afterward that I had even met Susan a few years earlier, too, and didn't know it.

Little did I know at the time that the young blonde girl I was drawn to in a picture during college would one day be my wife. In fact, I had been praying for her, her baby and a young fraternity pledge that came to me one night for counseling.

That's right, I found out *after* I met Susan that the biological father of our oldest child walked away from her before she was born, in spite of the advice I had given him.

It's an amazing love story that continues to be amazing even after 35 years of marriage.

21

JUDITH MCPHERSON
MIRACLE VOICE SAVES HER CHILD FROM DROWNING

Judith is a homemaker in Birmingham, Alabama

My husband Lane and I, with our three children, had spent the day on crystal-clear Smith Lake in northwest Alabama on our pontoon boat with our dear friends, David and Cindy, who also had three children. The two dads and four of the children, all fitted with life jackets, had jumped in and out of the lake all day while Cindy and I, along with our two nine-month old babies, enjoyed the activities. At the end of the day, the pontoon boat was parked back in its slip at the marina boat dock. The dock was approximately 100 feet long and contained about 20 slips. Our slip was third from the end farthest from the shore.

Cindy and I gathered up our kids and our things, then waited in the middle walkway beside the boat watching Lane and David finish packing. We then took off the kids' life jackets as the last things to be packed and locked up. At this point, our little five-year old, Brett, went to the very end of the dock to fish while Cindy and I were lightly chatting. We were almost ready to go to the cars after a wonderful day of play on the lake.

I heard my eight-year old friend's daughter, Rebecca, say to me, "Look at Drew!" I turned to see what our little three-year old son Andrew was doing. He wasn't on my right, so I turned to the left. He wasn't to my left or at the end of the dock. I made a 360° sweep and saw no sign of him! Cindy began to help me search.

I called for Andrew but there was no response. Great fear began to rise up within me. Had he wandered back to shore? He couldn't have gone far. There was no sign of him on the shore.

Time was becoming the enemy. I ran up and down the dock looking into each slip and between all of the boats. Because of the noise of the wood and steel docks creaking with the waves, neither of the dads, whose backs were turned to me, had heard me call out.

I began frantically running up and down the dock, holding a baby in my arms and calling out for Andrew. Dread swept over me. I ran back to our boat to look in the water surrounding it again. He was nowhere to be found and had no life jacket on.

I turned around to scan the water in the empty slip opposite ours. After a second, I saw him, near the end of the slip, the top of his head twelve inches below the water! The water was so clear I could see his blue eyes wide with panic, struggling frantically. He had been watching me the entire time but couldn't get his head above water!

I screamed like no husband ever wants to hear his wife scream, shoved the baby to Cindy, and dove into the slip. Lane came hurtling over the boat banisters and into the water after me. I shoved Andrew up to the surface, then Lane shoved me up to the surface where we were fished out by our friends. We all turned our attention to Andrew... He was essentially fine! We were all shaken. He had jumped into the water like he had been doing all day, but this time he hadn't been wearing a life jacket. How long had he been underwater? My search for him seemed like an eternity. How could he have held his breath for all that time?

Cindy said, "It's a good thing you thought to look for him." I replied, "I didn't. I was watching the guys. I wouldn't even have known to look for him if Rebecca hadn't said, 'Look at Drew.'"

"I didn't say anything," said Rebecca, "I was watching Dad and Mr. Lane." No one else was around me at the time. But SOMEONE had said the exact words, "Look at Drew." Otherwise, it would never have occurred to me to look for him.

We all realized something very Providential and miraculous had happened. There was absolutely no other explanation. The marina owner had said Andrew had jumped into 80-100 feet of water. Bodies are rarely recovered in Smith Lake due to depths that reach 200 feet.

Life Flight helicopters take 45 minutes to arrive in that remote area. God had literally saved Andrew's life by prompting me to start looking for him. The ride home was quiet and prayerful as the reality of what might have happened, and what actually did happen, began to sink in. God has purposes beyond our understanding and He alone gets the glory in this story.

〜

22

DIANE WITHERS

SAVED FROM TORNADOES

Diane resides in Alabama

Diane Withers and her daughter, Emily Smith, are friends from Birmingham. Several years ago, they were driving with Emily's two small children to Houston to visit Emily's brother. As they approached a town in Louisiana on Interstate-20, they both thought they were in Shreveport and began looking for their exit, Highway 171. They couldn't find the exit, and after getting confused and turned around, they pulled over to ask for directions.

To their embarrassment, Diane and Emily learned that they were actually in Monroe, Louisiana, 100 miles short of Shreveport! Instead of getting upset at this unplanned 20-minute delay, they laughed at their mistake and shrugged it off.

An hour and a half later as they cruised into Shreveport, they were still chuckling about their unplanned detour. They found the Highway 171 exit, and as they headed south, they noticed the skies getting darker. Unbeknownst to them, they were heading directly into a serious storm system.

A short time later, they were driving into a small, one-stop-light town on the Louisiana-Texas border called Logansport. They were enjoying worship music and singing together. The skies were overcast, but didn't seem too alarming.

All of a sudden, about 150 yards away, Emily saw a very tall dark-haired man in a trench coat suddenly appear on the road. (She later estimated that he would have had to stand somewhere near *eight feet tall.*) The man held out one of his hands, palm-outward, motioning them to stop. While still well over 100 yards away, Emily distinctly heard him say, with urgency and authority, "Stop!"

Being inside her car and with music playing, and being some distance from him, there was no way Emily could have heard him speak. Yet she clearly audibly heard him, and though he was nondescript in appearance, she immediately perceived that he was an angel. There was a sense of a "holy" fear at being near his presence. When she heard the word *Stop,* Emily immediately sensed the Holy Spirit saying, though not audibly, "Pull over and you'll be safe."

"Mom, we've got to pull over now!" Emily barked. Immediately, Diane pulled over and into the parking lot of a small trucker's diner that was directly beside them. "Don't you see it?" said Emily. "See what?" asked Diane. "Don't you see that angel? He said to stop." Diane never saw him but found herself suddenly facing the diner.

Still not sure what was meant by being "safe," Diane and Emily, along with Emily's two small children, entered the small metal and cement building that housed the diner. As they entered the restaurant, they were greeted by a waitress whose first response was, "Come on in. You'll be safe here."

Still confused about any danger, they sat down. The skies had quickly turned dark and hail began to fall. People from neighboring houses began running into the diner for protection; it turned out that the diner was actually the emergency storm shelter for the area. Uncertain what was happening, Emily called her sister, Allison, who lived in Birmingham. Allison was watching weather reports on TV and said that the entire area in western Louisiana around Emily and Diane was filled with tornadoes!

Then it happened.

The men and women in the diner heard a roar and felt an earth-shaking sound like a freight train. A tornado touched down several hundred yards from the diner and began coming directly toward them! Amazingly, as the tornado approached the building, it lifted into the air, jumped over the building, then touched down on the opposite side and continued on its destructive path.

Quiet came over the diner.

No one was hurt. As they cautiously went outside, there was destruction all around them. The area in front of the diner looked like a war zone – lights and trees were downed and phone poles had been snapped in two. Yet the only damage to the diner was a slightly bent metal sign!

As Diane and Emily returned to their car and continued their trip toward Houston, about twenty minutes beyond Logansport, they encountered another area filled with tornado damage along the high-way. They then began to reflect on their unexplainable 20-minute delay earlier in Monroe and realized that, without that delay, they likely would have been in the direct vicinity of these other tornadoes!

"If you make the Most High your dwelling—even the Lord, who is my refuge—then no harm will befall you, no disaster will come near your tent. For he will command his angels concerning you to guard you in all your ways; they will lift you up in their hands, so that you will not strike your foot against a stone." (Psalm 91:9-12)

23

DIANE WITHERS

MOTHER IS BORN DEAD THEN REVIVED TO LIFE

Diane resides in Alabama

Diane's mother, Mrs. Charlotte Hendon, was born in 1920 in Dustin, Oklahoma. Charlotte's family lived on a farm, and when Charlotte's mother, Edith White, was in labor, Edith became extremely ill. Two doctors in the community, Dr. Lett and Dr. Wallace, rushed to the home and discovered that Edith was actually in life-threatening danger from pregnancy complications.

As they frantically worked to save Edith's life, the baby (Diane's mother Charlotte) was delivered on the kitchen table only after being forcefully pulled out by forceps. The baby had no breath. She was born dead. The doctors took the baby and set it aside while they continued to try to save Edith's life.

Edith's mother, Mrs. Martha Foster (baby Charlotte's *grandmother*), picked up the dead baby and sat with her in a nearby rocking chair. Throughout the night as the doctors worked on Edith, Martha rocked the baby back and forth and prayed. For hours she prayed over the infant, massaging her little arms and legs. As the new day began to dawn, the baby's foot gave a kick!

Charlotte revived from death and became a live, healthy little girl! But what about brain damage? How can a baby's brain survive approximately six hours of no oxygen?

Simply, it can't.

The second miracle is that there was *no brain damage* or disability of any kind! Charlotte carried the scars from the forceps on both sides of her face for the rest of her life, but she was strong, healthy and smart. In high school she became her class valedictorian and went on to lead a full and normal life.

Interesting footnote: Martha Foster, Charlotte's grandmother who rocked and prayed for her throughout the night, was part of the Topeka Outpouring revival that began in 1901. This outpouring resulted in many miraculous works of the Holy Spirit. At Charlotte's birth, Dr. Wallace said she was the largest baby he had ever seen, and both doctors declared that her survival was a true miracle.

"I am the Lord, the God of all mankind. Is anything too difficult for me?" (*Jeremiah 32:27*)

~

24

TONY JEFFERS
FROM THE HARLEM GHETTOS TO IBM

Tony is a technology consultant with IBM who resides in Texas

I'm an African-American man who was born into poverty and grew up in the Harlem section of New York City. My mom was unsaved and was not a religious person. This story touches on a few of the many special things God has done in my life and how He has taken me from the streets of Harlem and promoted me through a career in high technology with Lockheed Martin and IBM.

As you'll learn in this story, I'm not very strong academically. The places I've been able to go in my career have been by God's grace and favor and supernatural interventions, not by my abilities.

As a child around nine years old, I began to gravitate toward church. No one else in my family had any interest in church, and there was no explainable reason why I was drawn there. I began going to church every week. Even as a child, I sensed more and more that I needed to get closer to God. No one was encouraging me to go. After observing this growing interest and desire in my heart to know God, Mom sat me down one day and shared this story of an encounter that happened when I was a baby.

"Tony, when you were about eight days old, you were in your bed. It was in the middle of the day when, all of a sudden, an angel appeared in the middle of the room and looked down at you. He said, 'I bless you in the Name of the Father, the Son, and the Holy Ghost.' I was terrified and so shaken up I couldn't say anything! The angel then looked at me and said, 'The boy will be taken care of.' Then he vanished."

After talking extensively with Mom, I'm convinced of the authenticity of this experience and that she was not under the influence of alcohol or drugs.

As I mentioned, I was not academically strong. The combination of living in poverty and being raised in a highly dysfunctional family greatly affected my education. In fact, it took all my efforts to just make C's in school. I consider myself below-average educationally. Throughout school, I was afraid of tests because I always failed them, and I couldn't write grammatically correctly until I was almost out of college.

As a result, I avoided things that required analytical reasoning. Para-doxically, I've spent most of my professional career in an arena requiring intensive analytical thinking.

When I received my first job in computers, I failed miserably. My boss would keep trying to fire me, but others would quit before he could do it, and he'd have to keep me on for a little bit longer. I was so frightened knowing that if I failed here, I'd quit altogether. It had taken me 13 job interviews to land this first computer job.

During this time, the Holy Spirit said, "Don't rely on yourself, rely on me." The Holy Spirit began to teach me how to analyze problems and work through them. I had no confidence in my abilities; because all of my work peers had strong educations and backgrounds, I had to rely in total trust on the Holy Spirit.

God has always allowed me to accelerate in job opportunities. In all but two-and-a-half years of my past 35 years in business, I have worked as an independent technology consultant. I've worked for companies such as DuPont, Lockheed, IBM, and the U.S. government. I've had to believe God for each contract. It has always been a faith walk. And interestingly, I've always been given the highest or second-highest income positions.

Here is an early example that has proven typical of how God has guided me through my professional career.

One day in 1984, I came home and told my wife, Christie, that God was calling me somewhere else. We had lived in Atlanta for five years, and my contract with Sarah Lee was almost completed. As usual, I had been spending daily time in prayer asking God to lead me.

Around the last week of the Sarah Lee contract, I received a call from a placement agency regarding an opportunity with DuPont Chemicals in Orange, Texas. After interviewing, I accepted the position, and Christie and I moved to the quiet town of Orange.

At DuPont, my boss was named John. If I may be candid, John was a stereotypical redneck-type person – he chewed tobacco, propped his feet up on his desk, and cussed a lot. John didn't have much respect for blacks, and he didn't know I was African-American when I was hired! You can imagine how this situation felt to both of us! To John's credit, he gave me a chance to prove myself, though I didn't receive much respect at first.

Shortly after starting with DuPont, all of the computerized personnel records for the Orange division went missing. This was a huge problem. There were a number of heated and frantic attempts to locate them. The files had been missing for several months when John called me into his office. "Tony, we've got a huge problem. We can't find the records. We've had our 20-year-experienced IBM guy and data guy looking for them. No one can find them." John was getting a lot of heat from his boss and had been given an ultimatum to "find those files or else!"

John gave me *several days* – not months – to solve the problem! I went upstairs to my office, closed the door and began praying. I had no idea what to do. God had always been good in giving me answers to problems. I prayed for 15 to 20 minutes. All of a sudden, He showed me a file name in my mind and revealed the answer.

The Holy Spirit directed me to go to the computer and pull up a certain file. I looked at the file name and noticed that the address was off by 8 bytes. I knew then what had happened and went downstairs to share this with the 20-year IBM guy. He listened to me then spurted out, "Oh my goodness, it's there! We can have those files back in about 15 minutes!"

In just over an hour of getting the assignment, the files that had been missing for two months were back! My three-month contract with DuPont] turned into five years. My relationship with John grew from me being perceived as a *liability* to being respected as a *savior*.

During my time at Dupont, there were other occasions when difficult things happened. Each time, the first thing John would do is ask, "Where's Tony?" God always gave me visions to solve the problem. His promptings were almost like a third person speaking – not audibly, but still clearly perceived.

Over my last 35 years in technology consulting, these types of experiences have happened numerous times. Many times, when starting a new contract with an employer, I would walk in and, within a few hours, could answer problems that had been issues for several months involving millions of dollars.

Interestingly, as Don and I were talking about these stories, I shared another example that had recently happened with my wife Christie.

Christie called me from our home while I was at work. She was frantic. "I've lost my keys; I've looked all over the place. You need to help me!" she said. Of course, I had no idea where her keys were. Then, while on the phone, I had another vision... "Okay, Christie, you were wearing something red..." "Oh my gosh," she exclaimed. She ran to the closet and found the red coat she had worn – and her keys were in the pocket!

Isaiah 30:21 "Your ears will hear a voice behind you saying, 'This is the way, walk in it.'"

～

25

DON CARMICHAEL

THE CALL TO CALIFORNIA CONFIRMED

Don is a businessman in Alabama

During the summer of 1987, I was invited to join the staff of a new ministry organization that was building an association of Christian business executives in the United States and Canada. This ministry was based in Rancho Bernardo, California, in the northern part of San Diego. Having just graduated from college, I was seeking a vocation with a meaningful purpose and felt this opportunity may be where God was leading me. But was it? How could I be sure?

Since the opportunity involved having to raise my own personal financial support, I sought God in prayer and asked Him to confirm this calling. Specifically, I asked Him to provide a Christian family with whom I could initially live at no cost and that would be located near my office. Ideally, I was hoping to find a situation where a child had grown and left home, leaving behind an empty and available bedroom.

About three months later, in Birmingham, my mom ran into a childhood friend named Wendy Jacoway at a dinner party. As they were catching up, Wendy explained that she now lived in San Diego. My mom mentioned my job opportunity there, and Wendy immediately offered to have me stay with them when I came.

Several days later, I called Wendy for more details.

She said, "We'd love to have you stay with us at no cost. Our son has moved out and is going to college. You'd be welcome to his bedroom..."

Knowing how large San Diego is and that it extends over 1,000 square miles (almost 50 miles from north to south and 25 miles from east to west), I was hesitant to accept her offer because of the likelihood that her home was far away from my office.

"What part of San Diego do you live in?" I asked.

"We live in Poway," she said.

As she described where it was in proximity to my office address in Rancho Bernardo, we both realized that, even though we were discussing two different communities, the office was in fact only a few blocks away!

In other words, out of a metro area of almost three million people and 1,000 square miles, *Wendy's house was less than half a mile from my office!*

The prayer was answered, the provision was given and the confirmation was made. I raised my financial support and moved from Birmingham to southern California to chart what would become the rest of my entire life's course.

~

26

DON CARMICHAEL

FIRST ANSWERED PRAYER – FROM FIGHTING TO FRIENDSHIP

Don is a businessman in Alabama

My first answered prayer occurred when I was a junior in high school, shortly after coming to Christ through the ministry of Young Life. The story actually began four years earlier with a boy named Kevin who had been bullying me.

In seventh grade, Kevin regularly harassed other boys and me at school. Trying to be obedient to my parents' to not fight, I refused to fight him – until one afternoon when he jumped me. My younger brother and I were riding home from school, doubled up on his bike, when we came to a four-way stop near our house.

Kevin, with his little sister watching, jumped into the middle of the street, grabbed our handlebars and threw us off. I landed on my feet, then Kevin began punching me in the face. We fought and tussled, and luckily I came out on top.

Four years later, we met again when Kevin began attending Mountain Brook High School. During those years apart, he had grown a lot bigger and tougher. He ran with a rough crowd and frequently got in fights. When he encountered me, it was payback time! For several months, he and another friend of his would harass me.

Then one day, Kevin and his friend caught me trying to get out of my car in the student parking lot. Classes had already started, and no one else was around. They began pounding on my window, kicking my door and shouting to get me to come out and fight. Each time I tried to open the door, they'd kick it shut again.

Angered and humiliated, I was shaking when they finally left. Instead of going to class, I drove to a remote place and prayed my first prayer since coming to Christ several months earlier. I asked God for protection and to somehow work in this situation to have Kevin stop.

Several weeks later on an early spring Friday evening, I pulled up to the home of a friend who lived in Kevin's neighborhood. I was on a date with my girlfriend Meghan. Before I could get out of my car, Kevin's car appeared, whipped around mine and cut perpendicularly in front of my parked car to block me in.

Exasperated and angry, I shouted, "That's it!" ... Kevin had now reached my open driver's window.

"Get out of the way, Kevin, let's settle this!"

But before I could get the door open, Kevin bent down by the open window and said, "Wait, wait!" Then this tough guy *apologized*!

"I've been thinking about the way I've been acting and treating you. I don't want to lose my friendship with Meghan. And I don't want to lose yours. I want to be friends. I'm really sorry for the other day – would you forgive me?" Dumbfounded, I shook his hand as he thrust it through the opening.

What caused Kevin to seek friendship instead of revenge? I believe only God could touch a hard heart in such a quick manner and turn a hostile enemy into a new friend.

~

27

BRITT HANCOCK

MIRACLE FEEDING OF MULTITUDES

Britt and his family served as missionaries in the rugged and remote mountains of central Mexico. He currently resides near Austin, Texas.

For eleven years, Britt's family has served as missionaries in some of the most remote and rugged areas of central Mexico. Most of their work is focused on bringing the Gospel of Jesus Christ to the indigenous Indians of the region who are direct Aztec descendants.

The entire region where they operate is dominated by witchcraft and demonic strongholds. Almost every family worships idols, which are usually given a prominent place of honor in their huts.

The natives in this region are desperately poor. Because there are so few options for medical care, the Hancocks see frequent miraculous healings. In fact, over 90% of the Indian converts to Christianity through their ministry have received or witnessed miraculous healings. Most of these healings occur over time, anywhere from several days to several months. Occasionally there are instantaneous healings.

The culture in their ministry region is very stratified. There are many who are very poor and a few who are wealthy. This is an account of a woman whose healing led to the salvations of many in her area, including one of the wealthiest families of the region.

This particular native woman had been suffering from jungle rot on one of her legs for over 20 years – a long open wound that stretched from her knee to her ankle. For years the doctors who had treated her were unable to heal it; it continuously oozed and would not close. She suffered terribly.

Some Indian converts of Britt's ministry had visited her three times over two weeks and, during this time, she was healed. The Lord closed up her wound and completely freed her from the pain and suffering it had caused for so many years.

Around this time, the Hancocks were trying to meet with new Indian converts who lived on the estate of a very wealthy rancher. The woman whose leg was healed was one of the ranch workers and lived on the ranch property. The owner told his staff not to let the mission-aries back on the property and denied future entry.

About nine months later, Britt's wife Audrey and he were hosting an American Thanksgiving dinner in their small home in the city where they lived. Audrey taught English to children in the area, including those of the upper class, as a way to become accepted members of the community. They had decided to use the Thanksgiving platform as a way to attract their parents so they could tell them about Jesus through the stories of how God met the needs of our early pilgrims.

During the evening, they shared the story about the healing of the Indian woman in the jungle and how God had healed her from years of suffering from jungle rot. One of the guests asked where this woman was located. It turned out that this woman was one of his workers! In fact, this young man of around 40 years old was the wealthy owner of the ranch from which they had been banned! This wealthy don shared how he had been paying this woman's medical bills for 20 years and how the doctors could never heal her. That night they talked for three hours about Jesus and how he desires to bring blessing and eternal life to those who would follow him.

The don came to one of the Hancock's services where the Holy Spirit touched his heart and he received Christ. Following this, he began to join Britt in witnessing to tribal natives. One day he offered to let them use his house to host an outreach event. The goal was to bring people to his home to learn about God.

In their culture, a host must provide food to his guests. To not provide food was considered a tremendous indignity and insult. They were expecting around eighty people to attend. The owner had prepared eight chickens and rice – and was worried whether it would be enough.

Then things got much worse – imagine his surprise and concern when they had 250 people show up! This created a potentially disastrous social problem for Britt's friend! He was not only facing terrible embarrassment, but even more importantly, this social insult might cause guests to be turned away from the Gospel message. Britt and Audrey asked a blessing over the food.

The don's wife told their fifteen staff not to eat. But how would this be enough to help? The staff began bringing out plates to the guests. There were so many. Yet there was still enough food to serve. Remarkably, the entire 250 guests made it through the meal with food on their plates. But then another huge problem began to develop. They wanted seconds!

The don's brother, who was serving in the kitchen, hoped against hope that no one would want seconds. How in the world would he be able to feed them? But they began to come... and come! As the staff shared later, the amount of food in the preparation pots never diminished. They would dip out food and dip out food, but the volume in the pots never went down.

In fact, the following day, Sunday, there was enough leftover chicken and rice to feed the don's family and staff. And on Monday there were still leftovers... and on Tuesday... and on Wednesday. By Thursday the don's wife called and said, "Do something with this food! Give it away or something... it's driving us crazy!"

God is an amazing lover of his people. How thrilled they all were and how they praised him for his miraculous provision of food — that eight chickens ultimately provided over 500 meals! And they praised him for bringing so many of those guests into an eternal relationship with him.

~

28

BRITT HANCOCK

ATTEMPTED KIDNAPPING AND SURVIVING A MACHINE GUN ATTACK

Britt and his family served as missionaries in the rugged and remote mountains of central Mexico. He currently resides near Austin, Texas.

One of the endemic problems in the central Mexican mountains where the Hancocks minister is kidnappings for ransom. This problem even reaches into the jungle regions where private ranches and estates are spread out.

The ranch owner mentioned in the Hancock's previous story had a first-hand experience with this problem. As a new believer in Christ, the ranch don began to grow in his faith. Yet even after the miraculous feeding of guests at his home, God had an even more dramatic experience in store that would even further deepen his faith.

The don was on his ranch when he was forcefully abducted by six armed men. Four of the men were armed with AK-47 machine guns, and two were armed with .45 caliber semiautomatic pistols.

They were escaping with the don into the jungle. He was praying fervently. A rope had been tied around his neck. The kidnappers were shoving him forward then jerking on the rope to keep him off-balance.

He was able at one point to get the rope off his head without being noticed. They came up to a fence. The four men with machine guns crossed over while the two men with pistols held their prisoner.

At this moment, the don explained that he felt courage and power come on him. He turned and disarmed one of his captors as easily as if he was a baby. With the pistol, the don shot three times in the air; he then dropped the gun and began to run!

When the initial struggle began, one of the men with a machine gun opened fire at him from just across the fence – at no further distance than about six feet – and emptied a 30-round clip. None of the bullets hit him, but they did hit his captors on each side!

A later investigation by area judicials revealed huge amounts of blood and found two banana clips taped together, opposite ended, with one clip empty. They found 30 empty brass shells laying on the ground. The don escaped without pursuit!

⌇

BRITT HANCOCK
SPEAKING IN STRANGE AND FOREIGN TONGUES

Britt and his family served as missionaries in the rugged and remote mountains of central Mexico. He currently resides near Austin, Texas.

I n the rugged and remote mountainous region of central Mexico where the Hancocks serve as missionaries, many miraculous signs and wonders occur among the native Indians. Most of the miracles are healing miracles, as health care is so limited in the area. But there are many other wondrous signs.

Most of the Indians are descended from Aztecs, and their language is an Aztec language greatly different from Spanish. One of the remarkable wonders occurs occasionally when these converted Indian brothers and sisters sing praises during worship services. On occasion, they have been filled with the Holy Spirit and have begun to sing praises in unknown tongues. Unknown to them, that is. Britt and his wife Audrey have witnessed these illiterate and isolated Indians singing in a foreign tongue – in English! They would actually be praising God and declaring his glories in perfect English!

YOUNG RUSSIAN AGNOSTIC
THE MIRACLE IN THE WAREHOUSE

The CoMission was an alliance of Christian ministries
focused on world evangelism

A ndrea Wolfe, on staff with the CoMission Ministry office in Raleigh, North Carolina, shared the following story about an angry young Russian agnostic who had a life-changing encounter with his Savior.

In the 1930's, Stalin ordered a purge of all Bibles and of all Christian believers. In Stavropol, Russia, this order was carried out with vengeance. Thousands of Bibles were confiscated, and multitudes of believers were sent to the gulags – prison camps where most died, unjustly condemned as "enemies of the state."

The CoMission once sent a ministry team to Stavropol. The city's history wasn't known at that time. But when the team was having difficulty getting Bibles shipped from Moscow to churches in Stavropol, someone mentioned the existence of a warehouse outside of town where these confiscated Bibles had been stored since Stalin's reign.

After the team had prayed extensively, one member finally mustered up the courage to go to the warehouse and ask the officials if the Bibles were still there. Sure enough, they were. Then the CoMissioners asked if the Bibles could be removed and distributed again to the people of Stavropol. The answer was "Yes!"

The next day, the CoMission team returned with a truck and several Russian laborers to help load the Bibles. One of these helpers was a young man – a skeptical, hostile agnostic collegian who had come only for the day's wages. As they were loading Bibles, one team member noticed that the young man had disappeared. Eventually they found him in a corner of the warehouse, head between his knees, weeping.

He had slipped away hoping to take a Bible for himself. What he did not know was that he was being pursued by his Savior. What he found rocked his unbelief.

The inside page of the Bible he picked up had the handwritten signature of his *own grandmother*! It had been her personal Bible. Out of the thousands of Bibles still left in that warehouse, he walked to this corner of the warehouse, stuck his hands into one of many boxes, and stole the very Bible belonging to his grandmother! Apparently, his grandmother was a woman, who throughout her entire life, was persecuted for her faith. Now approximately 50 years later, God honored the faithfulness of this grandmother and enabled her faith to have a lasting impact – on her own grandson!

No wonder he was weeping. God had powerfully and yet tenderly made Himself known to this young man. Such was his divinely appointed meeting with the sovereign Lord of the universe, his Savior who had drawn him to that very warehouse! Remember Jeremiah's words: "Can anyone hide in secret places so that I cannot see him?" declares the Lord.

31

TERRI HARTSELLE
A MIRACLE PHONE CALL FROM TEXAS

Terri* is a homemaker from Alabama

As a junior at Auburn University, I had returned several months prior from a summer job in Dallas, Texas where I had met a handsome young man named Eric. Our summer was full of times spent together enjoying walks, dinners and long conversations. We fell in love with each other, and life was wonderful!

However, toward the end of the summer, things changed. Through a series of events, our relationship began to strain, and on the last day, we got into a huge argument. Harsh words were said, and I left Dallas in tears with a broken heart.

A few months later, back at Auburn, I was involved in a weekly women's Bible study in my dorm and was sharing with my friends how unsettled I still was from the way things ended with Eric. I shared how I wished we could talk again to apologize and bring some closure. As we began praying together, I was moved as one of my friends prayed, "Lord, you know how much this broken relationship is hurting Terri, and how it's probably hurting Eric too. Please work out a way for them to talk."

I smiled as she prayed those words, then was shocked when she continued with an audacious request:

"... and Lord, because this is so important to Terri, would you have Eric call her as soon as we finish praying! Thank you."

I thought, wow, that's gutsy.

We wrapped up praying and someone said, "Amen." And then... it happened.

The phone began to ring!

Could it really be?.... Yes, it was Eric!

I hesitantly answered the phone as all my girlfriends watched this drama unfold with anticipation.

"Hey Terri, I've thought a lot about the way we left things last summer and just feel awful. I want to apologize. Can we talk?"

"Of course," I said. But I had to ask him, "What made you call right now at this specific moment?"

"I don't know," he said. "It was the strangest thing. My buddy and I were just driving down this road in Austin hanging out, and I had this deep sudden urge to call you. I couldn't shake it and told him, 'Quick, pull over to that phone booth! I need to call Terri...' So here we are."

* Pseudonym

32

DON CARMICHAEL

I BURNED YOUR HOUSE DOWN – YOU'RE WELCOME!

Don is a businessman in Alabama

My wife, Vicki, and I own a rental house that used to have three bedrooms, two baths and 1,700 square feet. That was before the fire.

Our company had leased the building for years as our office space. In 2010, we decided to move the company out of the house and rent it for extra income.

We both shared a deep sense that this house was to be used for more than just a source of investment income. We wanted to honor God with it and sensed that He wanted us to make it a place of refuge for a family in need. Fast forward a few months later, we rented the house to a man named Barry (pseudonym) and his family.

Barry was a Christian man whose business was struggling. He had to sell his home in an upscale gated neighborhood before moving into ours. As part of our due diligence, we reviewed financial and bank statements and saw that he had sufficient income to afford the rent.

We accepted his lease. Everything went fine... for a few months. Then slowly he began to get further and further behind.

It seemed that Barry had money to buy new TV's for every room, a new truck and a fancy surveillance system, but somehow couldn't afford the rent. During this time, we agreed to let Barry build out the unfinished basement into four bedrooms for his children and his elderly mother and father.

After several years of continuing to miss rents and promises that they would be caught up, the rent shortfalls had put us in a serious financial bind. The house needed repairs, a new AC system and a new roof, but we couldn't afford them. In fact, we were struggling to maintain our mortgage payments.

The end of the lease term, May 31st, was approaching. We were done, and I told Barry we would not renew the lease. I told him that we needed all unpaid rents caught up and needed him to vacate the house by that date.

At this point, Barry became adversarial, explained that he knew how to play the system, that he was definitely not going to move out by May 31st and that it would take six months to get him out. We were furious. Especially after all we had done for him and his family.

May 31st came. He didn't leave.

Two nights later, at just after midnight, I received a frantic call from Barry shouting, "Your house is on fire! Your house is on fire!"

Vicki and I jumped in the car and raced to the rental house. When we arrived, it was the main event on a slow night in Shelby County.

There were three fire trucks, three ambulances and at least two police vehicles, all with their emergency lights flashing. The roof of the house was engulfed in massive flames!

The fire was finally put out, and though the outer structure remained, the inside of the house was a total loss from water and smoke damage. Fortunately, Barry and his family all made it out unharmed, but they lost everything in the fire.

A week later, the fire inspector finished his investigation and showed us where the fire had originated in a bathroom light fixture. Squirrels in the attic had built their nest over a light fixture in the ceiling below where it would keep them warm. As they got bored, the squirrels chewed on the wires until... pop! A spark arced onto the dry nest and set the house on fire.

It all worked out as an amazing blessing from the Lord.

Our belligerent renters were permanently evicted about 48 hours after their lease expired. When our insurance agent arrived, he explained that we had a rental income replacement rider that provided guaranteed monthly rental income until the home could be rebuilt – a process that took a year.

And even better, we had a "replacement" insurance policy, which meant that the insurance company calculated our claim on replacing the house *as it was* prior to the fire. The claim was not calculated on a 3-bedroom/2-bathroom 1,700 square foot house, but because of the basement buildout, it was calculated on a 7-bedroom/2.5-bathroom 3,400 square foot house!

God incredibly and graciously rewarded us with a brand-new 7-bedroom, 3,400 square foot house, we believe, for so faithfully trying to bless our renters.

Though we had been soft landlords, we had good hearts – and we now have *new carpets, new appliances, two brand new HVAC systems,* a *new roof* and *four new bedrooms*!

We could just imagine God saying with a grin, "I burned your house down. You're welcome!"

∾

DR. DONALD CARMICHAEL
THE VISION OF THE TALL THIN SURGEON

Dr. Donald Carmichael is a retired vascular surgeon in Alabama

In the early 1970's, as the charismatic movement was in its early stages of becoming known to mainline churches, I as a physician wished to have a Sunday morning adult class on faith healing. In order to obtain a subject whom we could interview, I contacted an elderly friend who was part of the charismatic movement. She gave me the name of a man in her prayer group whom she thought would be willing to be interviewed.

I waited approximately three months to contact this man by phone. I called him one Saturday morning and asked if he would visit our Sunday class. He explained that he was very shy and did not think he could come. He did say that he would pray about it. He called back about two hours later and asked if I were a doctor. I said yes. He then said he would do whatever I desired of him.

We agreed to meet the following Saturday morning in my office to make further plans for the class. When we met, he told me of his experience of being a long-time Episcopal church member who had noted the observable joy of friends who had been "Baptized in the Holy Spirit." He shared that he desired to have the same experience and had asked God for this.

On a business trip, he heard some audible words which he believed to have been God speaking, and he pulled off to the side of the highway. He felt the presence of a very benevolent, warm spirit, which he felt to be the Holy Spirit. He then began to doubt his experience as being truly the Holy Spirit, and a sense of coldness, which he believed to have been a demonic spirit, came into the car. He commanded it to leave. At that time, the warm presence returned, and he was nonverbally told to confess all of his sins up to that point – which he did.

After this confession, he felt a love for all people, including two men who had wronged him prior and whom he had been unable to forgive up to that point. As he stood up to leave my office, he paused and said that he would tell me something he thought would be of interest to me.

He said that, at a meeting of his prayer group the previous week, the day before my initial call to him, a young lady who had recently become a member of the group told them of a strange vision she had seen. In this vision, she saw herself in a hospital looking into an operating room. There was a tall, thin man in his 30's dressed in white operating clothes who raised his hands and said, "Praise the Lord." At that moment, she realized his name was Dr. Carmichael, and the vision ended.

She shared the vision with the members of the prayer group who were meeting that Friday, again the night before I first contacted the man sharing the story. The prayer group said that they didn't understand the meaning of the vision, but did she know any Dr. Carmichaels? She said she did not, but that her mother had been a patient of a Dr. Carmichael – but he was an old man and not the young man in her vision (this would have been my father, Dr. John Carmichael).

The prayer group couldn't give an interpretation but would pray for Dr. Carmichael. The description of the doctor in the vision accurately fit me. I knew all of the Dr. Carmichaels in Alabama, and I was the only one who matched that description.

When considering the extraordinary timing of the vision, my phone call, our meeting, the clarity of my description she described and the number of Dr. Carmichaels in Alabama, I calculated that the odds of these happening in concert by coincidence were well beyond one-in-many-millions chances, and I concluded that God had indeed been actively involved in intervening in our situation.

34

HELEN MCEWEN

MY CANCER COMPLETELY DISAPPEARED

Helen and her husband Frank own a farm-to-table business providing farm-fresh eggs, beef and stone-ground grits to some of the finest restaurants in the country. She resides in Chelsea, Alabama.

About a week and a half before Thanksgiving in 2019, I called my friend, Dr. Cecelia Stradtman, to check on some stomach symptoms I'd been having that made me think that I had a prolapsed uterus. I went in to see Celia for an ultrasound - she didn't like what she saw, so she ordered a CT Scan for the following day. From the CT scan we discovered that I had a large tumor on my right ovary (the size of a softball), but what was of most concern was my abdominal wall. There were flecks of cancer all over it - "like a paintbrush" – indicating that the cancer had already spread into my abdomen.

I had a peace in my spirit from God like I've never had before. Beyond a natural sense of denial, I felt like the Lord was holding me in His lap. The next day I embarked on a three-day fresh juice fast/detox tea and tincture, and all weekend friends came to pray for me. It was a spiritually focused weekend like I've never quite had before.

My husband, Frank, reached out to our friend, Carolyn Shaw, who is a prayer warrior and sees visions often as she prays. He asked her to pray for me, and she said she would but assured Frank that I would be fine, as did all of the friends that came through over the weekend. Carolyn called me the next day and told me that while she was praying for me, the Lord gave her the same vision twice.

The following week I had surgery. I only had a 10% chance of having the full tumor removed. The doctor was pretty certain that he would only be able to biopsy the tumor, in which case he would treat me with chemo to try to shrink it before a followup operation. The waiting room was filled with friends and family. Frank was praying that the doctor would come out with a smile on his face.

My best friend Celia served as an assistant surgeon to Dr. Barnes, neither of whom were expecting to witness the unexplainable event that was about to take place. Instead of the tumor being 8 cm large, it was 13.5 cm and somehow removable. There were two nodes on my ligaments and some on the fatty layer that lays across the top of the colon. He was able to remove all of it.

The miracle came when he saw that *all of the abdominal wall cancer* had disappeared! Despite the evidence of dozens of visible spots of cancer shown on the CT scan, they were ALL gone! When I woke up in recovery, there sat Dr. Barnes with a big smile on his face telling me the good news. He said the cancer was all gone! I said, "Dr. Barnes, do you believe in the power of prayer? Do you know how many people have been praying for me this weekend and how many came to my house and laid hands on me and prayed for me?" (Luke, our son, mentioned that he was not exaggerating when he told me there were probably 1,000 people in Auburn praying for me through a social media blast).

Dr. Barnes just smiled and nodded. I dozed back off, and when I awoke the next time Celia was sitting by my bedside with the biggest grin ever on her face! She said, "You did great girlfriend! He cleaned you out!" I stayed in the hospital only two nights. Dr. Shah, who was on duty Thanksgiving morning sent me home (I was supposed to stay until Saturday). A friend, Karen Summers, called and said they were bringing Thanksgiving dinner for Frank and the boys while I slept peacefully on the couch. Such provision!

How did this happen? My doctor had no explanation, but without words I could tell that he knew it was divine intervention. I asked him at my two-week follow up if the black sand and water could have been what was on my abdominal wall.

He thought about it and said, "I don't know...I don't think so but you and I are (and he waved his hands and I finished what I thought he meant, "on the same spiritual plane?") and he said "yes, so at this point I would believe anything..." He reaffirmed that the abdominal wall cancer had disappeared. With no medical explanation.

∼

35

VICKI CARMICHAEL

THE VISION OF THE FIFTH MOUNTAIN

Vicki is an artist and homemaker in Birmingham, Alabama

A few years ago, our family attended a fun Family Weekend in north Georgia at Young Life's Sharp Top Cove camp property. There were close to a hundred families that came, kids and all. The weekend was full of different activities during the days and several talks in the evenings by a man named Ben Hand (pseudonym). During one of his talks, Ben described how he and his wife had been walking through four difficult challenges in their lives and some of the struggles related to them.

As he talked, I had a clear vision appear in my mind of Ben being carried in the strong but gentle grip of a giant eagle. That eagle was carrying him over the mountains below, but there were clearly five mountains in my vision, not four. I perceived that each mountain represented one of Ben's specific struggles.

After the end of his talk as people were gathered in conversations, my husband and I went up to him to thank him for his message.

As we were talking, I shared with Ben the vision I had seen and explained that I had seen *five* mountains, not four.

He reacted with mild shock! He shared how there actually was a *fifth* trial that he was walking through, but that he had not told anyone.

When I shared my interpretation of the vision, that I believed the eagle represented the Holy Spirit who had Ben firmly in his grip and was carrying him over these five mountains of trials, he was deeply moved and encouraged.

We were all excited by this sweet encouragement from the Lord and that He would care enough share it with us.

36

GLENN ESHELMAN
FROM HUMBLE DAIRY FARM TO GRAND SCALE IMPACT

*Glenn was raised on a humble dairy farm in Lancaster County, PA in the
heart of Amish country. He is the founder of Sight and Sound Theaters in
Strasburg PA. The 2,100-seat theater conducts grand-scale productions and
showcases a 300-foot stage (the largest in the U.S.) that wraps around the
audience on three sides. This stage design offers panoramic views of its
captivating sets, which feature structures up to four stories tall!*

*Each year, Sight & Sound hosts nearly 1.5 million people from around the
world who experience performances with live animals, state-of-the-art
lighting, technology and special effects that bring the Bible to life.
An estimated 24 million people have attended a performance since its
founding in 1976. Sight and Sound operates an identical theater in
Branson, MO and employs over 700 people.*

I was born in 1939 and raised in Lancaster County, Pennsylvania
in a very conservative home. My upbringing was humble, and
Christ was the center of my family. My uncle, great grandfathers and other ancestors were preachers in the Brethren Church,
and I was the fifth generation in my family to continue that honor.

The older I get and the more I look back to try to understand how all the extraordinary things in my life have happened, I have come to the simple conclusion that, in the natural, it was impossible. I am not a trained or highly educated person. God truly was the source of all giftings, guiding and blessings.

My education consisted of attending a small, one-room classroom for eight years. I was planning to be a dairy farmer. I was totally untrained in art and theater, and through many years of God's divine guiding, He took a simple dairy farmer and used him to bring world-class live epic shows of Biblical stories to over a million people a year.

The Bible says don't despise small beginnings. At the early age of around four or five, I found myself drawing various scenes I'd see during the day on the farm. I used crayons. In my conservative culture, art was not considered valuable, and we were never allowed to go to a theater. However, my parents saw my love and giftings for art. They gave me five tubes of paint and encouraged me. Thus began the story of how God took such little things – five tubes of paint – and turned them into a grand testimony of His impact and glory.

My wife Shirley is also from Lancaster County. We married when I was twenty and worked together, and I soon went into business painting Lancaster County farm scenes. Shirley handled the business end of it. I began taking reference photos for my paintings and then fell in love with photography.

In 1965, Shirley and I presented a program of scenic photography at a local church using a slide projector, a record player for musical underscore and a microphone for narration. The audience response was overwhelming. By the mid 1970s, we were taking these multi-media presentations to audiences around the country. In 1976, we built our first 750-seat theater in Lancaster County. Sight and Sound was officially born.

In 1987 we produced our first all live stage show titled *Behold the Lamb*. We soon outgrew that theater and built a 1,400-seat theater, which was one mile away. There we produced our first epic show, *Noah*, in 1995. Sight & Sound was rapidly growing. Then in 1997, that new theater and all of our sets burned in a tragic fire that destroyed everything. After discussing the difficulty of rebuilding, we realized this was more than a business or ministry. This was a high calling from God that He put on our lives many years earlier.

A fire can destroy a building but not a calling. We began the daunting task of building a new and much larger theater. Since our productions had grown to an epic size, we decided to step up our vision with a building to accommodate these grander shows. It would take a financial miracle to make it happen, and the miracle took place. We had around 300 employees at the time and were able to keep most of them employed through the rebuild.

One of the many amazing events that occurred was our water well. When we bought the 65 acres of land for the new theater, it had no public utilities. We had to develop our own water, sewer and power. We were ready to drill for water, but it was very difficult to find in that area. Our site director, Doyle, asked me to come down to help determine where to put the hole. We joined hands and prayed, and I sensed the Holy Spirit saying to place it on the hillside behind me. This did not make sense.

I walked up the hill and punched a hole in the ground where the Lord told me. It was very difficult getting the drilling rig there because of the slope. When we finally drilled, we hit a major water flow that was so strong we couldn't measure it! We drilled other wells on the property but could barely get any flows from them.

Of all the miracles I've seen in my life, and now as a man in his eighties looking back over a lifetime, I'm convinced that the greatest miracle of all is that Jesus saved my soul and cleansed me. If you don't know him, please get to know him. Your life now and in eternity will never be the same.

∾

37

MERIDA BROOKS

PINK DRESS IN THE CLOSET

*Merida is a native of Louisiana and served for over 30 years in Christian
school leadership and administration in Opelousas, Louisiana.
She currently resides in Birmingham, Alabama.*

In the early years of our marriage, Father God called my
husband and me to keep the principle of "praying in your closet," making your personal needs and desires known only to
Him. While reading through the book of Luke with our children, we
came across the "principle of the closet." After explaining the idea of
praying to God "in secret" for personal needs and desires, our two
daughters, Deborah and Lydia who were 4 and 5 at the time, jumped
up, each running to one of the two closets in our bedroom closing the
door behind them.

I quietly tiptoed to the doors thinking I would listen and "help" prove
God's character to the girls by aiding Him in meeting the requests in
their prayers. Without hearing anything I scurried away as I heard
one of them stirring. Once they both exited their respective closets, I

nonchalantly asked, "What did you request, girls?" In unison they promptly responded, "Mom, we can't tell you. It is supposed to be in secret." I was encouraged that they really understood the lesson, but also a little disappointed that I could not assist in seeing their prayers answered.

Several weeks later I received a call from one of the young, single ladies that joined us in our home each Friday night for Bible study. She asked if she and one of the other young women could come to the house to bring gifts for the girls. When they arrived, each had a wrapped box in their hands.

Deborah and Lydia's eyes widened as they saw the beautiful boxes and were each handed one to open. Both were thrilled and delighted, but it was Deborah who squealed, "It is true! Mama, it is true!" I could not yet see what was in her box as I asked, "Deborah, what is true?"

From the box she lifted the answer to her closet prayer. "I prayed for a pink dress with white lace!" In her hands was a beautiful light pink dress with an exquisite white lace collar.

God did not need the cunning of a concerned mother to prove His listening ear or His character and love for a five-year old. Father God heard Deborah's prayer in that closet, knew her heart's desire and revealed to her that He cares about the intimate details of her life.

Beginning to "see" God in the daily happenings of our lives is the first step in learning to "walk with God."

∽

38

MERIDA BROOKS
THE CASE OF THE MISSING KEYS

Merida is a native of Louisiana and served for over 30 years in Christian
school leadership and administration in Opelousas, Louisiana.
She currently resides in Birmingham, Alabama.

I received a call from a school in Bastrop, Louisiana, that was desperate to find a special education teacher to complete the last eight weeks of the school year. The original teacher of the 15-18-year-olds' class of emotionally and behavioral challenged boys had become sick in the fall. Two other "fill-in" teachers had already come and gone. Being an adventurous one, I said, "Yes," while praying to myself, "Lord, please help me with this!"

When I arrived, I was shown to a large classroom where you could actually see the ground below through gaps in the board floor. The assignment was teaching Shop! Yes, as in "building things!" Fortunately, the original teacher was present to orient me to the classroom and provide a run down on the students. His first advice was to lock up the power tools! Do leather working instead! Needless to say, I took that advice.

It was obvious he had care and concern for his former troubled students as he gave me insight into each of their specific needs. One student stood out in particular, whom we will call Demetrius. Demetrius apparently loved to steal things, and it was suggested that personal items be put in a secure place.

Things went along fine for several weeks. Nothing was missing... until one day when it was time to leave. My car keys were gone; nowhere to be found! Normally that might not be critical, but school was thirty miles from home. I was driving my mom's car, so she could not come for me. I was also the last one to leave school that day.

After searching for almost an hour, I began thinking I might be spending the night in my classroom. In desperation I prayed, "Father God, You know where Demetrius put my keys. Please show me."

Immediately my mind went to the birdhouse high in the corner of the classroom. I dismissed the thought primarily due to the daunting pile of lumber precariously placed underneath the birdhouse. I continued my search. The birdhouse continued to nag my mind. Finally, I thought about my prayer and I knew I was out of options. I needed to climb that stack! Climb I did, grasping onto shelves on the wall to steady myself as I ascended the mountain of lumber.

When I got close enough, I stretched out putting my hand into the birdhouse and felt something familiar! My keys were retrieved!

There was no way I ever would have considered that Demetrius had ascended that mountain of lumber, but Father God knew. He whispered it into my mind, and I spent that night at home.

39

TAMSIN EVANS
FROM MISCARRIAGE TO MIRACLE

*Tamsin is a native of York, England and is the
founder and CEO of Pure Creative Arts, a fine-arts theater company
that conducts inspirational dramas in public high schools in both
the UK and United States. She resides in New York City.*

Y*ou see an earthquake, I see an opening. You see a car wreck, I see
you hope again. You see a dead end, I see you breaking through.
You seem worn out; I'm making all things new. You see a lost
cause, I see a turnaround,*

*You see unwanted, I see you finally found. You see a locked cage, I see an
open door. You see dry bones; I see fire in your soul. (Lyrics by Christina
Boonstra, Tamsin Evans and Rael James)*

Becky ran to the bathroom. Loud sobs followed. I looked to where
she had been sitting and saw several large pools of blood on the sofa.
Then, on the floor, more blood than I had ever seen. My heart
dropped.

Becky was three months' pregnant at the time, and it looked like she was having a miscarriage right here in our home. Five minutes earlier, I had been talking about how God can turn impossible situations around. We were praying with our team of leaders for the impossible to happen through a youth project we were running. Now, right in the middle of it, this had happened?

I ran to the bathroom and found Becky crouching on the floor, surrounded by a growing pool of blood. There was no way the baby could still be alive. Holding her sobbing frame I began to say, "I'm sorry, I'm so sorry." Then I stopped myself. What was I saying? Did I believe what I had been saying just five minutes earlier? Did I believe God could turn this around?

I began to pray, speaking life to the baby inside Becky. I commanded a turnaround in Jesus' name, a reversal of what I could physically see. Faith began to rise up in me. I stopped worrying about whether my prayers of faith would hurt Becky if she lost the baby. Instead I put my focus on God and what He could do in this situation.

In the hall, Greg called Becky's husband Dave while someone else called an ambulance. The paramedics brought in a wheelchair to take Becky, still heavily bleeding, to the ambulance. In shock we stood at the door as the blue lights flashed down the street and out of sight. Closing the door left us to face the pools of blood that were still on the floor and the only thing we could do was pray.

"God, don't let it end like this, a cold room, pools of blood on the floor. We believe you for a turnaround. Let the baby live."

Later that night I received a text message from Becky. The doctors had told her that she had miscarried, and that they would do an ultrasound to confirm tomorrow. Then, a little later, they examined her and discovered her womb was still closed. How could it be? It didn't make any sense.

The next afternoon, she had an ultrasound and, to their amazement, the doctor found a heartbeat. The baby was alive! Becky was never given a medical explanation for what happened – that night we saw a miracle.

Eight months later, as I held Becky's healthy baby girl, Mia, in my arms, I felt God whisper to me:

"When the blood is on the floor, when it looks like it is all over, dead and gone, this is where I begin. I am the one who turns it around, who resurrects, who brings life from dead things. Believe."

(From the book, "Take Your Place," by Tamsin Evans)

40

BEATTY CARMICHAEL

MORE MIRACLE HEALINGS

Beatty is a businessman in Birmingham, Alabama

We were in a local Chick-fil-A restaurant line when a big man named Edward walked in on crutches with his family. I felt the Lord wanted me to pray for him. We got our food and sat in the booth next to him. I approached him and asked him what happened. Edward described that he had been cutting down trees; he turned his body to the right but his left leg got caught in the downed trees. When he turned his body, his left leg didn't turn and he ended up ripping the tendons in his left knee.

I asked him if he wanted God to heal him right now. He sheepishly grinned and said that he knew God could. To that I said that God would. On a scale of 1 to 10, his knee pain level was a 10 when he walked on it with crutches. I prayed for him twice and told him he'd walk out of the restaurant with the crutches in his hand. I had him check out his knee. He walked without the crutches, and then I had him do deep knee bends to really test it out. He did... and there was no pain. It blew his mind!

• MY FRIEND WEYMAN and I met a large man named Eddie at Walmart. The cartilage in his right knee was gone causing great bone-on-bone pain. He said that his knee pain level was usually at 9 or 10 on the scale, but only 5-6 on the scale today. Additionally, he had glaucoma in his left eye and described his vision in that eye like looking through very dirty windshield. Eddie said he experienced sharp pains when looking at bright lights through the glaucoma. We prayed for his cartilage to be rebuilt and, in testimony of it being rebuilt, have all pain disappear. All pain immediately left his knee. He could even do deep knee bends without pain. Then I prayed for his left eye. In two prayers the glaucoma disappeared, his eye became perfectly clear and all its pain disappeared.

• WE MET ROSE, an older lady walking her cart down an aisle, with three grandchildren and asked if we could pray for her. She had just lost her son a few months prior, who was the father of her 14-year old grandson with her. Rose said she had a blood clot in her left leg and had to wear compression hose for it. I prayed for her and had her test it out. When she did, she started to cry because she was walking almost without a limp. She could not feel the blood clot any more. But what seemed to impact her the most was that she said she had suffered terrible arthritis pain in her left knee and some in her right – and that most of that pain was gone. The grandchildren even noticed that she was walking so much better. So I prayed again, and all of the pain completely left. She was overwhelmed with the Lord's goodness in doing this for her. We stayed there talking about fifteen minutes. She shared that she was raising the grandchildren by herself, and I felt led to give her $250, which overwhelmed her even more.

41

SCOTT VAN DYKE

KILLER ROCK SLIDE ON MOUNT ARARAT

Scott is CEO of an international petroleum company in Houston, Texas

I n Scott's words... Following a vision the Lord gave to me, in 1983 I led a team of six climbers up the north face of Mount Ararat in search for the historic Noah's ark. It was late August when we embarked on our second Ararat ascent of the summer.

The climbers on my team were hand-selected. Walter Scott was one of my first choices, a native Houstonian and former roommate of mine when we attended Washington and Lee University. Walter is a very close friend, and I knew he would risk his life to save mine if I found myself in a desperate situation.

The other climbers on the team, in addition to myself, were Jessie Reinhardt (our official photographer who now lives in Jerusalem and has a ministry to Jews), Grant Richards (our team medic and a geologist), Steve Nootenboom (aka "Spider Man," an expert mountaineer who could climb anything), and Umut Ekol (team translator and a Turkish medical student – who after our climb became the youngest Turk to ever reach the summit of Mount Ararat).

Our late-August climb focused on the Ahora Gorge on the north face of Mount Ararat. This gorge is an extremely large canyon that averages 1 - 1.5 times the depth of the Grand Canyon and has a similar width across. Our mission was to find the remnants of Noah's ark. As a volcanic mountain, in general Mount Ararat is extremely porous. The large ice caps among the upper heights of the mountain send their melted water through the top of the mountain, down through internal water channels, and out through springs along the base of the mountain. There is very little water between the bottom and the top, and as a result, there was very little water to be found on climbs.

Another challenge rock climbers face on Mount Ararat is the extensive prevalence of loose rocks and a terrain notorious for rock slides. The mountain is filled with thousands of small gullies and narrow ridges. As we climbed, we tried to stay on ridges and avoid the gullies as much as possible to limit the danger from rockslides.

On this particular morning, as we were preparing to begin our climb, we had finished breakfast and had dropped down into a gully area where there was a small stream. We were still high up on the mountain at about 11,000 feet. Everyone was moving about the stream, brushing their teeth and filling canteens. Because gullies are more susceptible to rock slides, I told my teammates to get back up to the ridge as soon as they finished at the stream. One by one they moved up to the ridge and waited.

I was the last person to finish and was the only one at the creek. I had just stood up when I heard and felt something whiz past my head. As I turned my head, I saw a rock about the size of a football that had just missed my head by inches! As that rock hit and bounced down the hill, I immediately turned and looked uphill to see what else was coming. To my great alarm, I saw an entire pack of large rocks hurtling toward me!

With the main body of the slide already approaching me, I didn't have time to run for cover. (There were some large boulders nearby that could have shielded me, but there was no time to reach them.)

With no options of avoiding the impending onslaught of rocks, I found myself instinctively raising my hands over my head, palms forward. I screamed at the top of my lungs, "In the name of Jesus you will not hurt me!" Being a trained geologist, as soon as I said this, I thought, "They can't hear you; they're just rocks!"

At this point the slide began to go over me. Every rock that came toward me went either over me or around me. Those that were in motion actually deviated from their natural paths of inertia – in mid-air. As I held my hands above my head, I knew that I knew – with my whole being I knew – that I was not supposed to lower my hands.

The rockslide seemed to last forever. I don't know if it lasted a minute or several minutes. When the slide was over and the dust cleared, I looked up at my buddies on the ridge to the right. They were standing there looking at me with awe. I looked down at my hands, knelt down to pray, and thanked God for saving my life.

From the ridge, my friends said they saw the slide coming toward me. Immediately they knew I had no chance of surviving it. They broke out in spontaneous prayer praying for my safety. I was obscured from view during the slide, but throughout this whole time my friends prayed.

I want you to know that this miraculous encounter was not merited in any way by who I am or by anything I had done. It's all about God's work. When a person is absolutely dependent on God, he realizes he's helpless. When we totally surrender our lives and are willing to say, "God, do with me as you desire," we can't help but have total awe, gratitude and humility when we see Him work on our behalf out of His great love and grace.

"If you make the Most High your dwelling – even the Lord, who is my refuge – then no harm will befall you, no disaster will come near your tent. For he will command his angels concerning you to guard you in all your ways; they will lift you up in their hands, so that you will not strike your foot against a stone." (Psalm 91:9-12)

42

DON CARMICHAEL

GRANDMA SALLY SPIRIT

Don is a businessman in Alabama

In 1996, my family and I returned to Birmingham and temporarily moved into my parents' home. While there, my two oldest daughters shared the same bedroom that my youngest brother, Troy, had lived in years earlier while a teenager.

As a young teenager before he came to know Christ, Troy had become involved in the occultic game of *Dungeons and Dragons* (and even prior to this he had seen and felt the presence of demons). As a result, he made himself vulnerable to attacks from demonic spirits. In fact, on one occasion in that same room, a demon had manifested itself to him.

Fast forwarding twenty years, my oldest daughters now occupied the same bedroom. Caroline, the younger of the two, was three years old when she started talking to "Grandma Sally." Of course, no one could see Grandma Sally. Was this just an innocent imaginary friend, or was Grandma Sally something more sinister?

One day Caroline came downstairs and announced that she had been with Grandma Sally, and that Grandma Sally had told her to "murder all the puppies" and "cut their heads off!" This was not normal language from a sweet and very innocent three-year old!

Recognizing that Grandma Sally was actually a spirit that was seeking to deceive our daughter, Vicki and I sat down with Caroline and Katheryn, three and five years old respectively, and taught them about spiritual warfare – about "good" angels and "bad" angels. We taught them that the 'bad' angels were called demons, and that they pretended to be nice to trick them and ultimately hurt them.

We explained that Grandma Sally was really a bad angel, and that the only way to make her go away was to use the all-powerful Name of Jesus. We explained that there was no name or authority greater than that of Jesus, and that the bad angels fear him and have to obey him. The next time Grandma Sally came to her, Caroline was to say, "Grandma Sally, in the Name of Jesus, go away!"

For several weeks we coached the girls in this response, and they knew exactly what to say.

Not many weeks afterward, Caroline told us that Grandma Sally had come back to her.

"What did you do?" we asked.

"I told her, 'Grandma Sally, go away in Jesus' name' – and she did."

That was the last time Grandma Sally appeared, and she never returned.

~

43

DON CARMICHAEL
BUSINESS ADVERSARIES RECONCILED

Don is a businessman in Alabama

In 1995, I experienced a fallout with two business partners that led to the most difficult season of life for Vicki and me. We had a substantial amount of money invested in the company and lost it all due to the unethical actions of one of the two partners.

The other partner's name was Hal. Hal had been one of the wealthiest men in Oklahoma through oil and gas and was a man I had greatly admired. But the fallout brought animosity. We engaged in legal action against the partners and won a settlement. The unethical partner left the company, and Hal and I parted ways.

Ten years later, I discovered that we were scheduled to exhibit at the same trade show at the Hyatt Convention Center in Dallas, Texas. We were now business competitors, and the last person I wanted to see was this big man and his bigger-than-life presence.

When I arrived in Dallas, one of my reps met me at the airport and drove me to the hotel. On the way, he got lost, and we spent an extra fifteen minutes meandering until we found the Hyatt. As we drove up to the hotel, there was only one person in sight on the long sidewalk that ran the entire length of the hotel – Hal! My heart began racing. I gave a partial wave as we drove past, but he didn't see me.

Twenty minutes later, I was behind the hotel unloading our exhibit gear at the hotel's loading dock. As I walked inside and turned the corner, there was, again, only one person in sight – Hal. I turned a corner and literally almost ran into him!

"Hello, Hal," I said calmly, my adrenaline pumping. "Hello, Don." We shook hands. Awkward. Again, nonchalantly as possible, I said, "It's good to see you. Maybe we can grab some coffee and catch up..." Why in the heck did I say that! He said that sounded good.

The next day during a lull on the exhibit floor, Hal and I left our booths and found a restaurant upstairs where we could talk. As we began catching up about each other's families and learning of the events in each others' lives during the past ten years, we both began to ask the other for forgiveness for things said and done during the legal fight. This big man came to tears several times as we each repented and sought forgiveness. I found myself brought to tears several times too. Our lunch turned into a three-hour visit between old friends. It was a special and deeply heart-moving encounter.

The next day during another lull, Hal came over to my booth and invited me to a coffee break – and we enjoyed another visit that was filled with grace, kindness and forgiveness.

As I left Dallas, I was filled with praise over what God had done in orchestrating the crossing of our paths and the restoring of a relationship and a friendship.

To top it off, several months later, I received two handsome and valuable shotguns from Hal as friendship gifts. One he'd had for almost 30 years. The other was almost 60 years old and had belonged to his father who had died when Hal was just 13 years old. What kind gifts and tangible examples of a restored relationship.

44

EMILY SMITH
SAVED FROM A BIKERS' ASSAULT

Emily is a homemaker who resides in Alabama

Emily was on staff with YWAM (Youth With A Mission) when she was 18 years old. She was living with a host family in a small west-Texas town outside of Midland, Texas.

One Friday evening in the summer, she and two YWAM girlfriends were driving back to their host home after an event in Midland. As they drove through a mostly-Mexican town on the way, they pulled over at a laundry mat just before sundown to wash their clothes. It was in a rough-looking area and seemed to be somewhat isolated.

As they were just starting to dry their loads, a group of eight tough-looking Mexican bikers pulled up to the laundry mat, saw the girls inside, and began leering at them through the windows. They were snickering and looking menacingly at the girls.

The woman who worked there, after seeing the bikers pull up and apparently recognizing them, left the counter, walked into a back room and quickly closed the door, leaving the girls alone! All three girls began to get nervous and prayed for safety. "Lord, we thank you for your protection. We don't know what's going on but thank you that you're sending your angels and protection."

Following that short prayer, a small Mexican man appeared outside the door near the bikers. He was wearing a string bolo tie, black cowboy hat, black jeans and boots, and a fancy bright green and turquoise Mexican cowboy shirt. He had blue eyes, a reddish complexion as if he spent a lot of time in the sun and seemed to be about 50 years old.

Emily and one of the other two girls grabbed their clothes, still damp, and prepared to quickly leave. The third girl, Shelly, said she wanted to wait five more minutes to get her clothes drier. The little man walked into the laundry mat, came up to her and said, "Now Shelly, you need to hurry up!" Shelly was shocked. This little Mexican man had been standing outside, yet he called her by name – and didn't speak with any accent.

The man went back outside and stood on the boardwalk past the doorway. The girls grabbed their clothes and walked out the door and went straight to their car. The man stood between them and the bikers, who were armed with chains and covered with tattoos. The bikers stood and silently watched the girls as they left.

As soon as the girls started their car and began to pull away, they turned around to see what was going to happen to the man – but he was gone! He had completely disappeared from sight. There was nowhere he could have hidden, as the entire area was wide open and without any cover!

When the girls got back to their host family's home, they shared about their biker encounter with their hosts. When the girls described where they'd been, they were told that it was an extremely dangerous and lawless area where rapes occurred almost nightly. They were protected through prayer by a small warrior from heaven.

45

BOB TIEDE

REVIVAL ON THE FARM

Bob is a leadership development executive.
He resides in Dallas, Texas.

T his story occurred in the early 1900's on a farm near Freeman, South Dakota. The story was told to my friend Bob Tiede by his grandfather, Jacob Walters.

One hot summer day just after the turn of the last century, a preacher on horseback stopped by the Walters farm. He explained that he was a traveling evangelist and asked Bob's grandparents, Jacob and Mary, if they'd be willing to help host an evangelistic crusade in several weeks on their farm. They said yes. "Do you all have a big tent or know where you can get one?" he asked. Jacob and Mary said no, and that they had no idea where they could find one.

According to Bob's grandfather, the evangelist then said, "Let's just pray that God would provide a tent." So they prayed. He gave a date of several weeks later when he'd be back for the revival meetings. The evangelist then mounted his horse, waived goodbye and began heading toward another community.

The next morning, a distant neighbor of the Walters, whom they barely knew and who lived some miles away, showed up in their barnyard. "Jacob and Mary," he said. "Do you know of anyone who might have need of a big tent?" "Well how big?" asked Jacob. "A really *big* one," said the neighbor.

Bob's grandparents then shared the story of the traveling evangelist from the previous day. After hearing this, the neighbor said, "This tent has been in my hayloft for years. If you need it, come get it." Jacob and Mary sensed God was preparing to do something special.

At the appointed time several weeks later, the evangelist returned to the Walters farm and, beneath that big tent, began his revival meetings. He preached for several weeks there on the farm. During the entire first week, not one decision was made by any visitor to trust Christ. Bob's grandfather shared how he and his wife would observe the evangelist walking beneath the moonlight after each meeting, praying for the lost. During the second week, the dam burst. One night, a large number of people (possibly 15-30, as this was a sparsely populated farming community) came forward to dedicate their lives to Christ, and others did the same on the following nights.

A week later, Jacob was in Freeman on business. One of the town's business owners said, "Jake, I hear a lot of people have been meetin' on your farm." Jacob replied, "How'd you know, I haven't seen you there." The business owner responded by saying, "Well, a lot of folks are coming in and paying their bills." Lives had indeed been touched.

In 1980, almost eight decades later, Bob returned to this small farming community in South Dakota. He had come back for his grandfather Jacob's funeral. During the service, Bob shared this story and how that revival had touched the community of Freeman. Following the service, an elderly gentleman came up to him and said, "Bob, I want to verify that story for you. You see, I was there. I was one of those who, on one of those summer nights, came forward to dedicate my life to Christ at the tent meeting on your grandfather's farm."

46

MERIDA BROOKS

THE CRASHING TREE

*Merida is a native of Louisiana and served for over 30 years in Christian
school leadership and administration in Opelousas, Louisiana.
She currently resides in Birmingham, Alabama.*

My husband was the pastor of a small church, and I
worked as an administrator/teacher at a local Christian
school. We lived on very modest income, and with five
children, we often depended directly upon Father God to meet our
needs. We had the blessing of walking in obedience to God's call
upon our lives, and that gave us many accounts and stories of seeing
and experiencing the loving care of our Heavenly Father in a
direct way.

One evening our family gathered in the den for prayer. As we moved
around the room, each shared their prayer concerns and requests
about relationships, happenings at school, concerns for others,
upcoming events, and individual needs. Prom dresses for the two

oldest daughters; tennis shoes for our son and other practical needs were listed. We began to lift our prayers and needs to Father God.

Our time of prayer was interrupted by a loud crashing sound. We all jumped up to rush outside. We did not observe anything that could have been the source of the noise until we made our way around the side of the house!

There, we saw our large twisted, cypress tree draped over our outdoor storage building, yet "caught" in the y-shaped trunk of one of the oak trees. A portion of the roof was collapsed but the tree had stopped six inches short of crushing our lawn tractor.

There was no storm, no mighty wind blowing... nothing! Our son immediately responded in frustration, "That's great! Here we are praying for things we need and now this happens." Ahhh, but with different eyes and a longer-range perspective I replied in a gleeful voice, "John, insurance money!"

Insurance money was forthcoming. My husband and a friend were able to repair the shed. Prom dresses were supplied and new tennis shoes purchased with left over insurance funds.

Once again, our children experienced the faithful care of our God. Now, as adults, they, too, are teaching their children to watch for God in their daily life happenings, which is the first step to a life of "walking" with Him.

~

47

RAY ANDERSON

A VISION LEADS INTO THE BUSH

Ray is a carpenter in Florida.
He served many years as a missionary in Kenya.

"For the past three months, long lines of people waited outside the gate of the mission station. We had been involved in food relief for the area trying to help the best we could. Twice a week we would drive to the town of Lugari, some forty miles away, to load our Rover with white corn for distribution. This particular day I overloaded the Rover and broke a spring. That afternoon I arrived back at the mission and unloaded the corn and spent the rest of the afternoon repairing the spring while Ommani continued with the rationing. It had been a long difficult day. By late afternoon, I was ready to call it a day.

As I made my way to the screened-in porch and headed toward my favorite homemade chair, the house girl handed me a cup of hot Kenyan tea and fresh roasted peanuts. With the smell of dinner being prepared, I began reflecting on the day's activities. I think there was a song entitled 'Forty Miles of Bad Road,' or something like that. Whoever wrote that song must have lived in Kenya. In a Land Rover, forty miles of rough road can feel like a week's ride. I was worn out and looked forward to a hot shower and bed.

To say I fell asleep sitting in that chair would not be correct, but for a very brief period of time, in a vision, I could see myself traveling down a bush path. It seemed I was desperate to find a certain grass hut. Darkness had engulfed me as I searched the hillside for that hut. I could hear wailing in my mind and knew someone had died.

I found myself in the rear of the Land Rover clawing feverishly at the bags of corn and bananas and crying out to God, 'God why did you bring me here? I drove all this way into the night and now I can go no further. It is dark and I don't know where to leave the corn!' I saw the flicker of fire coming from within the grass hut. That seemed unusual because once locked inside at night a Kenyan family usually will not venture outside.

Standing in the glow of the fire, I could make out a young man and his wife who appeared to be holding a small baby that I sensed had just died. My heart was saddened as I heard the young man express his anger at God. 'If there is a God, where is He? If He is a God of love, why has he done this terrible thing? If God is real, come to me tonight.'

With a gentle nudge, Ommani reminded me that dinner was being served. I regained alertness and sat down to eat but could not get those images out of my mind. 'Ommani, get Samuel to help load four bags of corn into the Rover as well as bananas. I will load up a bag of sugar and tea. We must deliver it tonight to a needy family.' Ommani was as worn out as I, but he motioned for Samuel to help load the Rover. With the loading completed, we began an evening journey to we knew not where.

By now the very things were happening just as I had seen in my mind. We drove into the bush. For no apparent reason, I would turn left. Without thinking I would turn right. I'm sure Ommani thought I knew where I was going. How wrong he was! In fact, I was beginning to think that I would not be able find my way back out of the bush when I finally approached the hillside just as the sun dropped behind the hill. The bush path abruptly ended, but not before the truck dropped the front wheels over an embankment. I quickly slammed the Rover into reverse. It was going nowhere. We were hung up.

The Rover was resting on the frame. To keep trying to reverse only made matters worse. Maybe by unloading the corn and bananas we could get enough weight off the axle to allow the frame to come up just enough to free us. Frustrated, I grabbed the bags of corn and single-handedly dragged them to the side as anger mounted. I could not figure out why I thought I could help someone way out here.

I scanned the darkened hillside for some help. Possibly we would be able to push the truck off 'high center' and get out of there. As the African night began to spread its cover over the area, I noticed a small flicker of firelight coming from within a grass house, and someone was running down the hill toward us. The young man stopped short of us "Wewe ni Munguu? (Are you God?) Wewe ni Munguu?" He asked for the second time. By now I had forgotten our situation and tried to explain to the young man that I was not God; however, I believed God had sent me here.

He invited us into his mud hut and asked his wife to serve us some tea, which they had just prepared. Her eyes cast a reluctant glance as she obeyed and motioned for us to sit at the table.

I shared with the family the purpose of our coming and gave them the heavy sacks of corn, bananas, sugar and tea. They looked at each other in disbelief as the young man began speaking. 'My name is Musa, (Moses in English) this is my wife Lydia. You are our guest. The tea we gave you with sugar was our very last. It will be days before we can get more. Our baby is very sick. His name is Gamalio, (Gabriel) our firstborn, and he is suffering from malnutrition. His mother can no longer give milk. We believe he is going to die. This very night I cried out to the heavens asking if there really was a God, why did He cause us to suffer? Lydia is a Christian. I told her if the God she serves is real, let Him come tonight. We heard the Land Rover coming through the bush. I could not believe that God would drive a Land Rover. But now I see how He uses His servants to do His work. Thank you for your help. I now know that God hears our cries and now I want to become a Christian. Please sir; can I become a servant for God? How can I know Him?'

My mind dug up a story in the Word of God about Moses the deliverer. I could not help but think how God had prepared Moses on the backside of the desert for a greater service, and that God would do the same for this young man on the back side of his desert of opposition. Ommani explained how he must be "born again" by asking God to forgive him of his sins and for Jesus Christ to become Lord of his life. Kneeling on the cow dung floor that night, Musa became a Christian.

By now his brothers had joined us, also asking God to come into their lives. We now had enough people who could help us get the truck out of the ditch. We loaded the family into the Rover and took them to the hospital for treatment for Gabriel. Somehow the dinner we missed at the mission station had lost priority.

Another two months passed before the rains finally came. The supplies given to Musa and Lydia were enough to carry them through until they could harvest their beans and peanuts. By this time Gabriel was a live wire of activity.

Without knowledge of the Bible, Musa started a church at his house. I asked Ommani how he was able to preach a sermon without knowledge of the Word of God. "It is quite simple." He replied. "You must understand that the fifteen or twenty people that come to his church every Sunday hear the same message over and over again. It is the message of his salvation and how God hears the prayers of people and supplies their needs. Every Sunday a new person comes to receive Christ. In the market, he will preach the same message you preached some time ago almost word for word. Many times he will repeat his conversion experience. The congregation knows in advance when he is going to share that experience. He starts his message the same way. 'I did not find God, He found me, in a very bad situation. We all go through life searching for God, when in fact it is we who are lost and need to be found. We seek the wrong things and ask the wrong people this question. "Are you God? But know this, when we ask God who He is, He will answer...'

This happened in 1972. Musa was able to attend Bible School, raise eight children, marry them off and inherit many grandchildren as proof of God's provision. And what about the baby God rescued during that famine? Gabriel found a wife, attended college in the states, and began a teaching career in the capitol city of Nairobi Kenya. The last I heard he received his doctorate with plans of teaching at the university."

48

JOSH SMITH

HEALING OF TOMMY'S KNEE

Josh stocks convenience stores and resides in Alabama

At a gathering of friends on New Year's Eve, 2007, I met a new friend named Tommy. Tommy had hurt his knee several months earlier and was wearing a knee brace. He was in a lot of pain. Apparently, it had been the final insult of a very tough year in his life. Tommy is a believer in Christ. That night several of his friends and I were praying for him and for the healing of his knee. As I was praying, I silently sought a very specific word for him and felt the Lord say, "Josh, you're making this too hard."

About this time some others with us and I had placed our hands on his knee. I asked, "Tommy, do you believe the Lord wants to heal you?'" He said yes. "Okay, I'm going to pray for you, and afterwards I want you to test it out to see if it's still injured... Father, we know that your word in Isaiah 53 says that by your stripes we are healed. Lord, you have already provided the healing. I pray you would enter this healing into Tommy's knee. Amen."

I then said, "Now Tommy, will you try to make it hurt?" Tommy tried moving his knee. He extended his leg slowly, then began bringing it back and forth, faster and faster. Feeling no pain, he then took off his brace and began to run around the room praising God. Though he had been suffering great pain for several months at this point, He was instantly completely healed through his faith and a simple prayer!

49

JOSH SMITH
MIRACLES IN A CONVENIENCE STORE

Josh stocks convenience stores and resides in Alabama

In January of 2008 I went into a convenience store in Falkville, Alabama. I work for a retail products company where my job is stocking convenience stores with a variety of different products. This story I share humbly as the power displayed in it is from God, Jehovah Raphe (The Lord who Heals); I was simply an instrument in his hands.

When I entered this particular store, a BP station with a convenience store, there was only one person on duty, a woman who was suffering with flu. She was holding her head in her hands and felt miserable. I knew her casually from my route and said, "Mrs. Barbara, if you weren't sick any more would you know it?" She said she certainly would.

"May I pray for you?" I asked. She was very willing, almost excited, and said yes. I prayed a very simple prayer along the lines of, "Jesus, would you please heal Mrs. Barbara? Amen." I asked her, "Mrs. Barbara, do you feel better?" She said no.

"That's okay. Let me tell you a story." At this moment, before telling her the story, an older customer named Freddy, who was about seventy years old, came in on crutches. I also knew him casually from previous times in the store and believe he was a former Marine. I told him, "Freddy, you're just in time to hear a great story." From previous conversations with him, I knew he was on crutches because of a decades-long struggle with neuropathy and suffered greatly with extreme and debilitating pain in both feet.

I told both Mrs. Barbara and Freddy about my friend Tommy's knee injury and how God had miraculously healed him several weeks ago on New Year's Eve. As I began sharing how Tommy had reacted after being healed, Mrs. Barbara began kicking out her knees. For years she had also suffered with terrible pain in her knees but had not told me. From my previous experience with Tommy a few weeks earlier and a deepened faith in God's power and his radical desire to heal, I felt the confidence to boldly represent him to Barbara and Freddy. As we talked, I felt a sense of special anointing come over me.

I turned to Freddy. "Freddy, I want to pray for your feet." Some customers began coming in, and Freddy and I were in their way. I said, "Let's step outside." We went out the door, and Freddy sat down in a chair in front of the building. "Freddy, I'm going to pray for you, and when I finish praying, I want you to try doing things you can't normally do." He agreed.

"Lord, thank you that we don't have to beg you for healing..." I prayed a simple prayer and laid my hands on Freddy's feet. I felt no warmth, no goose bumps, nor any physical signs or feelings that people often say accompany healings. "Okay Freddy, can you stand without your crutches?"

"I'm afraid my knee will go out of socket," he said. This was news to me. I didn't know he had knee problems. "Well Freddy, let's knock that out too!" I said.

"Well pray for my hip as long as you're at it," Freddy said.

I didn't know anything about his hip problem, but with confidence in the power of the Great Physician Jesus, I prayed for Freddy's hip, his knees, and his feet. "Okay, now stand up."

Freddy, without his crutches, began to stand up very slowly and tentatively. "Freddy, can you make your feet hurt?"

"I'm trying to make my knee pop out of socket, but it won't," he said. Freddy then stood up, lifted up on his toes, and remarked that he couldn't do that before. He began walking around and around with bigger steps. Eventually he sat down and popped back up. He did this several times with growing excitement!

As I went back into the store, I learned that Mrs. Barbara had been healed from her knee problems and the flu. Both she and Freddy had been instantly healed! I found out later that, a year earlier, Barbara's son had died, and in her anger and grief, she had turned her back on God. Yet as a result of his loving healing work in her body, she was able to experience healing in her walk with the Lord.

~

50

DENNY DURON

BROKEN DOWN TRUCK PROVIDES FOOD FOR 3,000 KIDS

Denny is a pastor in Louisiana

I have a dear friend named Rick Berlin. Rick has a heart for inner-city children. Some years ago, we felt called by God to begin a camp for these high-risk children in our city of Shreveport. We called it the Winner's Circle; it involved conducting one-week sessions throughout each summer. To our amazement, we had over 1,500 kids sign up for our first camp – and we had committed to providing lunch for all of them! This was a huge problem.

It was the Thursday before these camps were to begin on Monday when Rick came to me and said, "Denny, I've got everything organized. Now I would like to go buy the food for the camp."

"We don't have any money for food!" I said. Rick replied, "What are we going to do? We've got to feed those children twice on Monday."

"We have to pray," I said. So we prayed. The prayer was something like, "Lord, you know we can't feed these children, but we know you can. Please provide what we need."

The next day I received a phone call from downtown. The unfamiliar voice on the line asked, "Mr. Duron? I heard you need some food. Well, I've got a truckload of food here and I would like to bring it over to you if you've got a freezer."

I said, "Yes sir, I've got a freezer here on my back porch, and I'll clean it out to make room for the food you have. I certainly appreciate it."

He laughed and said, "You don't understand. It's not a pickup truck load. This is an eighteen-wheeler full of food!" He continued, "The strangest thing just happened. This truck driver was going through Shreveport on I-20, and his truck just turned over. Nobody's hurt, thank God, but we've got to do something with this food before it spoils. Do you have a big freezer?"

"Yes sir I have a freezer." The gentleman on the other end asked, "What's the address?" I said, "I don't know, but if you will call me back in five minutes I'll tell you the address."

I knew that God was not going to provide an eighteen-wheeler full of food if he didn't intend to provide a freezer to put it in. My office manager, June Barnhill, called a local freezer company, and the owner of the company donated the space when he heard what we wanted to do with the food. When the man called back, I gave him the address and said, "Sir, before you get off the phone, would you mind telling me what's in the truck?"

I was willing to take whatever it was – even if it was all fish and vegetables. My new friend replied, "Oh, we've got steak fingers, chicken fingers, french fries..." All the food that kids love! We fed our camps for two years from that truckload of food.

I'm convinced Almighty God loves to see hearts joined together in a fervor of faith. He loves the son or daughter who is willing to say, it's okay. We will just pray it in.

∿

51

BEKELE SHANKO

WITCH DOCTOR'S FAMILY SET FREE FROM DARKNESS

Bekele is the Director of Cru for Southern and Eastern Africa

My name is Bekele Shanko. I was born and raised in a rural village in Ethiopia. This is a story of how God rescued my father, and ultimately my family, from darkness and demonic oppression. In 1973, when this story occurred, most of the villages in the area were still under spiritual darkness; the hearts and minds of many villagers were under the control of the devil.

My father, whose name is Shanko and who is presently 81 years old, had three wives. He belonged to a clan that was led by a powerful man, a witch doctor. This man lived on the top of a mountain teaming with snakes and all kinds of wild animals. The witch doctor had mighty demonic powers – he could bring and stop rains, and when he cursed people, they would be immediately struck down with sickness or death.

My father was one of the closest allies and an assistant to this powerful witch doctor. But this relationship came with a huge price. Dad was living under a huge demonic influence, drinking alcohol, smoking, beating his wives, and trying to obey numerous demonic instructions. He was required to wake up every morning at 5:00 am to begin drinking and smoking. He was not allowed to eat various kinds of food. When a meal was served at home the first portion was given to demons.

Whenever my father failed to observe some of the instructions, the demons would become very angry with him. Frequently they would afflict his children – my brothers and sisters. As a result, and due to related circumstances, eight of my brothers and sisters died from demonic attacks. I am the second of the five children in my family who survived childhood. I was not given a name until I was about four years old because my parents were not sure whether I would survive. At the age of four they named me Bekele, meaning "he has germinated."

When I was five years old, Almighty God sent two angels to visit my father. The angels explained the reality of heaven and hell. Through a vivid dream, they gave my dad a vision, in essence a tour, of heaven and hell. After this tour they asked him, "Where would you like to be?" Of course, my dad replied, "Heaven!" Then the angels told him, "We will send two men to you. They will come and explain to you how you can get to heaven. You must listen to them."

Two days later, two men, who had been converted to Christianity about a week earlier, came to see Dad and my family. These two men explained to us about the love of God, forgiveness, and new life through Jesus Christ. Light, salvation, and freedom came upon my family that day!

Two days later, my father was walking alongside a river in the village tending cattle. He had never been to school and was unable to read or write. But while tending the cows he found a New Testament Bible lying on the ground. It was a mystery where the Bible had come from. My dad picked up the book and started opening the pages. All of a sudden, he heard a voice say, "This is my Word."

Confused over where the voice had originated, he went and sat down under the shade of a tree and asked God to open his eyes so he could read and understand the book. Immediately, he experienced a miracle and was able to read the Bible! Shanko has been unable to read any other materials since this time – books, brochures, newspapers – even to this day, but he has always, over the past 35 years, miraculously been able to read and understand the Bible.

That evening, he invited the whole village to gather and hear his story. The whole village, approximately 400 people, came together. My father told them about what had happened to him and asked them to deny Satan and believe in Jesus Christ. Knowing the manner of man my father had been and now hearing this testimony, the entire village came to Christ. That was the beginning of the spread of heavenly fire in those villages.

God has since been demonstrating His power in those villages and in the whole country of Ethiopia. Since the time of this story 35 years ago, the number of Christian evangelicals in Ethiopia has grown from a small 200,000 to over 12 million. As a result of my dad's conversion and my coming to Christ, I now give leadership to the ministries of Campus Crusade for Christ (Cru) throughout Southern and Eastern Africa, consisting of 23 countries.

52

JAMES KARMICH

RESCUED FROM TROUBLE

James is a businessman who resides in Alabama

During college while on summer break, I traveled to Raleigh, North Carolina with my mom to keep her company while she worked on a family project. After several days, she decided to extend her time there. Being somewhat bored and ready to return to my friends, we decided that I would return to Birmingham, my home, via a Greyhound bus.

During the first leg of the trip from Raleigh to Atlanta, I sat next to a girl about my age. She was very pretty and, from all appearances, probably promiscuous. I was a young Christian, and as we continued to talk and flirt, I was convicted that this was a person I shouldn't be flirting with.

It was dark when we rolled into downtown Atlanta. We had an hour layover between our arrival and the connection with the bus to Birmingham. For a young white college student, spending an hour in the downtown Atlanta Greyhound bus terminal – at night – was crazy! There were all types of people there, most of whom looked sketchy and rough.

As the time drew closer to head to Birmingham, I felt more and more convicted of my intent to sit with this girl and make out during the long bus ride. I knew it would happen and was excited about it.

About ten minutes before boarding the Birmingham bus, in the midst of hundreds of bustling down-and-out-looking people, I suddenly saw a friend I knew from church in Birmingham named Mike. Mike was blind, in his thirties, and was making his way by my seat.

It was such an unusual thing to recognize someone there that I found myself blurting out, without thinking, "Hey Mike, what are you doing here?!"

No sooner had I spoken those words than a nicely dressed older couple behind Mike exclaimed, "Praise the Lord! We asked God to send someone who could sit with Mike and take care of him all the way to Birmingham! And here you are, the answer to our prayers!"

Noooooo!....

I was stunned. I was caught and had no way out. And as frustrated as I was about being stuck with Mike and not getting to enjoy my female friend, I was even more excited over how I knew God was present, at work in my life, and was protecting me for my good!

So how did the rest of the trip turn out? The girl sat a few seats behind me with some other guy, and they fooled around in the dark all the way to Birmingham.

I, on the other hand, sat with Mike and listened to him talk and tell non-stop terrible corny jokes! I was drained when I arrived home, but I praised God for moving in my life and protecting me from who knows what.

～

EVIDENCE OF AN INTELLIGENT CREATOR

SECTION 2

As demonstrated through the preceding "super-natural" experiences of just a handful of people, there must be some force or power at work beyond our physical four-dimensioned world (length, width, height, and time). What is this force? How does an intelligent and rational person explain the *super*-natural elements of the stories you just read? Can they be explained by simple physical explanations such as hallucinations, extra-sensory perception or mere coincidences? Or is there a greater force at work? If so, what is this force?

In this section, you will be presented with amazing facts and insights into the brilliant designs of life, nature, and the universe. I hope you will be awed, as I am, by the spectacular engineering behind even the simplest forms of life.

The Big Question is: *Where did all of this magnificent and intricate engineering and design come from*? Can it be rationally explained by random evolution, or is there a greater power at work?

A few years ago, my wife Vicki and I went to encourage some friends whose three-month old infant had just been rushed to the hospital. We arrived at the Birmingham Children's Hospital emergency room and were directed to a screened-off patient room. Inside the room were our friends, Matt and Allison, their baby Reagan, and several friends of the family.

For about twenty minutes we talked and joined them in praying for Reagan. Allison, who was sitting on the bed, continued to hold him in her arms while we prayed. Reagan looked completely healthy. He had a peaceful countenance and seemed to be sleeping quietly. Then it dawned on Vicki and me: Reagan wasn't asleep; he was dead.

We were shocked.

Reagan had died prior to our arrival. Matt and Allison were dealing with this devastating situation with composure and calmness. They believed, like we did, that God was very able to bring Reagan back to life if He so chose. I was moved by their reaction to this tragedy, which was the result of SIDS.

But at a much deeper level, I was confronted by a profound question: What is *life*?

Reagan's little body was fully formed. He had everything necessary for life in place – all of his arms, legs, bones, heart, eyes, mouth, brain, skin – he had it all, but where had his actual *life* gone? His brain was fully formed with all of its nerves and synapses in place, but where was the *life force* that ran it all? The body was complete, but *life* was absent.

Put yourself in this scene. You are holding a perfectly formed marvel of creation – a human body. It's complete in every physical detail. But without *life,* it is just a bag of flesh. What, then, is *life,* and where did it come from? Is life simply a consciousness that results from electricity firing in our brain – or is it more?

And even if one describes life as mere *consciousness*, where did this consciousness come from? Could it really have evolved from simple non-living physical matter?

THE BIG QUESTION

As you look at the following remarkable examples of life, nature and the universe, continue to ask yourself The Big Question:

Where did all of this come from?
Did it just happen by accident, or is there an intentional *intelligent design* behind the order?

Regarding this Big Question of where did life come from, there are only two options:

Random Evolution or *Intelligent Design*

It must be one or the other.

Which option has the most *evidence* to support it?

Which one requires the most *faith* to believe it?

I encourage you to consider the evidence, and let it lead you to your own intelligent and informed conclusions. In this section we will look at the following information on the Master Designs of Life:

A. The Spectacular Designs of the Human Body

B. The Spectacular Designs of Animals

C. The Spectacular Designs of Plants, Earth, and Nature

D. The Spectacular Designs of The World and Universe

E. Significant Problems With Darwinian Evolution Theory

THE ROLEX WATCH

Think about this. You are hiking on a high mountain trail in the Swiss Alps. There's nothing around you but beautiful trees, majestic mountains and blue skies. As you're walking along, there on the path in front of you is a beautiful Rolex watch.

You pick up the watch and notice that it has a precision-tooled stainless-steel wrist band with yellow gold-brushed band insets and an 18-karat gold unidirectional rotating bezel mounted on a blue ceramic bezel insert. You notice that a clear sapphire crystal covers the dial. The watch face is blue stainless steel and contains precisely engraved lettering, numbers and the name 'Rolex.' On the side of the face is a cut-out date box that is displayed under a cyclops lens in the crystal.

As you study it more, you realize that it is a self-winding automatic chronometer that has a 48-hour power reserve, beats at 28,800 vibrations per hour and contains 31 jewels. Beneath the face you can glimpse the extraordinary movement of tiny springs, sprockets and gear systems.

You're standing there holding this watch in your hands. I know what you're thinking, in fact I can read your mind.

"Cool! Look at this example of random evolutionary perfection! All of these natural ingredients in the steel, gold, sapphire crystals, screws, springs, sprockets and gears just happened to form into these precise shapes, bump into each other and make this perfect time-keeping chronometer, leaving it here on the ground for me to find!"

If I were standing next to you holding this watch and you heard me make this statement, you'd think I was an idiot.

In fact, you'd know I was! Why?

Because there's NO WAY those elements could have *randomly* been created, brought together with such precision and performed such an intentionally designed function. It defies scientific laws, laws of probability and simple rational logic.

Yet, this is Darwinian evolution theory.

Darwinian evolution, at its core, is based on a complete denial of any possibility of an *outside force* of intelligent design.

Instead, it claims that all of life and what we observe in nature is simply the product of random evolutionary processes that began as *simple cells* bumping into each other, like the parts of the Rolex watch, and growing incredibly *more complex* over time into *marvels* of complex cellular structures, genetic formations and physiological systems.

The Theory of Evolution sounds plausible on the surface, but as you read further, you'll discover that there are devastating fundamental challenges to this theory that make it scientifically impossible.

As you begin to read the following examples of the brilliant designs observed in the human body, think about the Rolex Watch.

As beautiful and perfected a chronometer as that Rolex watch is, it doesn't come close to competing with the exquisite designs, precision engineering and functional capabilities you can observe in your very own body!

SPECTACULAR DESIGNS OF THE HUMAN BODY

Your Body Has Been Brilliantly Designed!

The fact that you are able to read this text is in itself remarkable. Think about it. You are instantaneously receiving and processing images of text through your eyes and into your brain at a processing speed many times faster than that of early computers.

Your brain is translating this text into words and, simultaneously, translating these words into the English language. From the English language, your brain is able to interpret meaning and simultaneously understand the thoughts and concepts presented in this book. Wow!

Were you designed by an intelligent Creator, or did you simply evolve to such complex capabilities from primitive algae?

Here we go. Get ready to be awed by the following facts about your own human body!

~

Your Busy Brain

Not only is the human brain one of the most complex structures in the universe, it is also one of the busiest. In fact, during intense concentration, your brain can burn as many calories as your muscles do during exercise. Intense thinking can literally be as exhausting as a physical workout!

More than 100,000 chemical reactions go on in your brain every *second*. Among the brain's many jobs is chemist. The brain produces more than 50 psychoactive drugs. Some of these drugs are associated with memory, others with intelligence, still others are sedatives. Endorphin is the brain's painkiller, and it's three times more potent than morphine. Serotonin is produced by the brain to help keep our moods under control. The brain also makes dopamine. Dopamine makes people more talkative and excitable. Another hormone regulates hunger.

The brain is also a radio transmitter that sends out measurable electrical wave signals. In fact, the brain continues to send these signals for as long as 37 hours after death! Our tiny knowledge of the brain is enough to show us that this incredible organ is no accident. The brain is much more than a powerful testimony to our Creator. Its incredible powers convince us that there is much more to existence than this material life.

(Reference: McCutcheon, Marc. 1989. The Compass in Your Nose... Los Angeles: Jeremy P. Tarcher, Inc. Credit: Creation Moments, Inc. Foley, MN. 800-422-4253. www.creationmoments.com.)

Your 20-Watt Brain

Picture in your mind the sight and sounds of popcorn popping. As you picture the popping becoming more frantic as the hopper fills up, do you begin to smell the popcorn in your mind? That's part of the wonder of the brain. The brain can not only store words and ideas, but sights, sounds and even smells.

The average person's memory is able to retain about 100 billion bits of information – the information found in 500 sets of printed encyclopedias. Think about this: *where are your memories actually located? How are they kept in your mind?* The summation of all of your knowledge of math, ideas, facts, people, concepts – and those memories from your earliest childhood and past vacations – are stored in the chemicals of your brain. How is this even possible?

To use computer language, the brain is not only a place where information is stored, it is also an information processor. Yet, it only weighs a little less than four pounds and uses about 20 watts of energy. It took decades before modern computers began to even approach such efficiency.

Research has shown that the more you use part of your brain, the larger that part becomes – just like building muscles. And if you don't use part of your brain, it starts shrinking. Few of us have developed our ability to memorize things to any great extent.

To show you what can be done, in May of 1974, a Burmese man recited from memory 16,000 pages from a Buddhist religious text! What are you doing with that marvelous organ, the brain that your Creator gave to you?

(Credit: Creation Moments, Inc. Foley, MN. 800-422-4253 www. creationmoments.com.)

~

Your Brain Versus a Computer

There is a huge difference between computers and the human brain. While the computer transmits information using electricity, the brain communicates its information using powerful chemicals. Let's say that you accidentally touch a hot pan on the stove. In less than two-tenths of a second, millions of reactions take place while your brain performs a huge number of operations. As a result, you pull your hand away from the hot pan – very quickly!

Depending on the need, your brain uses many different chemicals for communication within itself and with the nerves that connect your brain to the rest of your body. As mentioned above, over 50 such chemicals have so far been identified. And it has been learned that many of these chemicals work in combination with each other so that over 800 different messages are possible.

Nor does the brain have set paths for messages, like a computer. In fact, it has been compared to a chemical soup rather than the circuits of a computer. In addition, the information paths in your brain change based on experience. If we can use computer language for a moment, the brain writes its own programs doing many complicated tasks of which you and I are never aware. We know that the computer is the product of careful design, but to argue that the much more complex human brain is an accident of nature makes very little sense.

(Reference: Hammer, Signe. 1986. How Does it Work? Science Digest, June. p. 45. Credit: Creation Moments, Inc. Foley, MN. 800-422-4253. www. creationmoments.com.)

Your Amazing Eye

The human eye – your eye – has 266 identifiable characteristics and is the most data-rich physical structure in your body. Far more individually specific than your fingerprint, there is only one in the 10^{75} chance two peoples' irises will match. On your retina, there are 120 million rods and 7 million cones. Your rods accomplish night vision, dim vision and peripheral vision. Your cones are for color and detail. Each eye has *over a million* nerve fibers that electrically connect its photoreceptors to the visual center of your brain.

Consider how brilliantly engineered your eyes are. Light rays enter your eye through the cornea, the clear front 'window' of your eye. The cornea's refractive power bends the light rays in such a way that they pass freely through the pupil, the opening in the center of the iris through which light enters the eye.

The iris works like a shutter in a camera. It enlarges and shrinks, depending on how much light is entering the eye. After passing through the iris, the light rays pass thru the eye's natural crystalline lens. This clear, flexible structure works like the lens in a camera, shortening and lengthening its width with rapid speed in order to focus light rays properly.

Light rays come to a sharp focusing point on the retina. The retina functions much like the film in a camera. It is responsible for capturing all of the light rays, processing them into light impulses through *millions* of tiny nerve endings, then sending these light impulses through over a *million nerve* fibers to the optic nerve. The optic nerve then carries these electrical impulses to the brain where the image is translated.

Not only do you have eyes for seeing, but specifically two eyes that allow you to perceive both distance and depth.

Today's computer digital landscape is very impressive but cannot compete with the retina's real-time performance. Computer scientists have learned that, before one single image is ever sent to your brain, each cell of your retina must perform a huge number of calculations. Each second, the cells in your retina perform over *one-hundred-million* calculations! And you thought you couldn't do math!

What's more, these are not simple calculations. Dr. Joseph Calkins, professor of ophthalmology at Johns Hopkins University, estimates that the fastest computers in the world just a few years ago would take hundreds of times longer to do what the cells in your retina do each second.

When you consider the extraordinary complexity and numbers of components that have to work perfectly in order for you to see – the millions of rods, cones and cells that work in conjunction with your flexible lenses, your retina that captures images then sends them through millions of nerve fibers to your optic nerve then ultimately to your brain for interpretation – it's easy to understand why so many scientists who study eyesight cannot accept the idea that the eye simply evolved through random evolution.

(Credit: Creation Moments, Inc. Foley, MN. 800-422-4253. www.creation-moments.com. Article Reference. National Keratoconus Foundation, https:// nkcf.org/about-keratoconus/how-the-human-eye-works/*)*

Your Fail-Safe Heart

Your heart will beat some 100,000 times today. That's over 36 million heartbeats a year and over 2.5 billion times in a 70-year life span. A healthy heart ticks along, producing beat after beat, whether you are awake or asleep. If you become more active, your heart increases its beating to meet the increasing needs of the rest of your body.

Doctors tell us that it's amazing how few of these beats are faulty. They say that it's perfectly normal for even a healthy heart to produce an occasional irregular heartbeat. Sometimes an irregular heartbeat is noticeable, but most often it's not. Doctors say that when your heart seems to skip a beat, it has really only beat prematurely. The premature beat leaves a pause before the next regular beat, making it feel as if your heart skipped a beat.

The clockwork precision of the heart's continuing beats is controlled by a built-in pacemaker. The pacemaker, called the sinus node, is a group of cells in the heart's upper right chamber. However, research has shown that every cell in the heart is able to send the electrical signal needed to produce a heartbeat if the sinus node fails. Using highly complex computer models, medical researchers have only begun to understand the electrical action within the heart. The human heart is much more than a pump, as once believed. It is also a computer and a regulator. Every beat of this wondrously designed biological machine glorifies the Creator who made it.

(Reference: Offbeat. Fairview Healthwise. P. 7. I. Peterson. 1983. "A Computer's Heart: Simulating the Heart's Electrical System." Science News, Mar. 19. P. 183. Credit: Creation Moments, Inc. Foley, MN. 800-422-4253. www.creationmoments.com. Article Reference)

~

Your Immune System:
100,000 Active Sentries

Despite the fact that most microorganisms are necessary and good, throughout our lifetimes, each of us encounters tens of thousands of different infectious bacteria, viruses, fungi and parasites. Even more remarkable is the fact that most of the time our immune systems disable these potentially lethal invaders before we ever show any symptoms of infection.

At any given time, more than 100,000 unique sentries posted throughout your body identify invaders, sound the alarm and even issue specific chemical instructions for their destruction. These sentries may also be thought of as tiny doctors who identify a potential illness, discover the cure and apply it even before the infection gets underway.

The immune system has puzzled scientists. Researchers know that our bodies do not keep a set of genetic blueprints for these sentries, which are called B Cells. How then does our body make these sentries or develop the genetic information necessary to disable invaders? Researchers have learned that the body has a small library of DNA fragments that are continually being shuffled into new patterns so that the body is almost instantly ready for any invader. The fact that even medical researchers are in awe over the design of our immune system verifies what the Bible says, "I am fearfully and wonderfully made."

(Reference: "MIT Researchers Isolate Master Builder's Disease-Fighting Gene." Minneapolis Star Tribune, Dec. 22, 1989. p. 2. Credit: Creation Moments, Inc. Foley, MN. 800-422-4253. www.creationmoments.com.)

Your Body's Self-Repair Ability

Your body has an incredible system for repairing itself. Suppose you accidentally cut your finger. Almost instantly, a series of precisely ordered steps begins to repair your finger. First, the bleeding must be stopped. While the scab is forming over the surface of the wound, the blood below is making another kind of clot out of blood platelets and protein.

With the bleeding stopped, your body increases the flow of blood enriched with white blood cells. These white cells not only search out and kill germs, but they also clean the wound of damaged cellular tissue. Skin cells begin to increase the rate at which they make new cells in order to bridge the cut with new skin. Underneath, cells called fibroblasts fill the wound to strengthen the tender new tissue and contract to pull the wound closed.

Now blood vessels and nerves complete their repairs as the fibroblasts position themselves along the lines of stress to prevent future damage. The intelligence in the order in which the steps to healing take place, as well as the advanced biochemistry involved in making those steps happen, makes the healing of a cut finger practically a miraculous event.

(Reference: "How a Wound Mends." Science Digest Magazine, May 1983. P. 86. Credit: Creation Moments, Inc. Foley, MN. 800-422-4253. www. creationmoments.com.)

❧

Your Multi-Purpose Nose

Some things don't have to be understood to be appreciated. You can enjoy the smell of dinner cooking or the scent of a rose without any idea of how your nose works. And believe it or not, you're doing just as well as the most brilliant biologist. While your nose knows how it works, science cannot explain just how we sense scents. It is known that inside our noses, behind the bridge of the nose, are cells that can sense smell. These cells are able to detect and identify airborne molecules from an open rose or a cooking roast. But no one knows just how these cells turn those molecules into the sense of smell that we experience.

To make matters more complicated, the sense of smell is one of our most complex senses. A single seemingly simple odor may contain more than 1,000 different chemicals. One sniff is likely to start activity all over the brain. Scientists have proven what experience has already shown most of us – a smell can also trigger emotions and memories, depending on an experience related to that smell.

In addition, your sense of smell is linked to your sense of taste, which is why food can seem to be tasteless when you have a head cold. A sense of smell has saved countless lives and brought joy and pleasure to all but those few whose sense of smell has malfunctioned. Yet it is so complex that modern science doesn't know how it works.[1]

Less well known is the fact that our noses are directly hooked into the brain's limbic system. Your limbic system plays a central role in generating and controlling emotions. This connection is why certain scents are able to raise strong emotional memories in us.

The nose is also designed to filter out and catch particles that should not get into the lungs. When a particle is taken in through your nose, it is snagged by the tissue inside your nasal passages. Next, this highly special tissue decides the best method of getting rid of the foreign material. If a bacterium is identified, your nose begins to run, bathing the bacteria in powerful chemicals that dissolve most of them. If the particles are larger, like pollen, a sneeze is triggered that can eject particles at speeds of up to 100 miles per hour!

Perhaps the most remarkable ability associated with the nose is its ability to act as a compass. Human beings have a small amount of iron in the bony part of the nose between the eyes. Scientific experiments have shown that we have an ability, while not foolproof, that can sense direction that is influenced by magnetic fields![2]

The Creator made your multi-purpose nose not only to bring pleasure to life but also to help protect your health.

[1]*(REFERENCE: Reyneri, Adriana. 1984. "The Nose Knows, But Science Doesn't." Science 84, September. p. 26.)*

[2]*(REFERENCE: McCutcheon, Marc. 1989. The Compass in Your Nose... Los Angeles: Jeremy P. Tarcher, Inc. p. 95. Credit: Creation Moments, Inc. Foley, MN. 800-422-4253. www.creationmoments.com.)*

~

Your Insulated Ears

While most people believe that our sense of sight is the highest of the senses and the most marvelous in design, our sense of hearing is no less marvelous. When a sound strikes your ear, your eardrum vibrates with the sound waves, fast or slow, soft or hard. These variations in vibration provide us with important information about the nature of the sound we are hearing. Some sounds produce a vibration in the eardrum as small as a billionth of a centimeter – only one-tenth the diameter of a hydrogen atom!

There are three tiny bones in the middle ear called the hammer, anvil and stirrup. They pick up the vibrations from the eardrum, amplify them and send them on to the cochlea. The cochlea is filled with about 25,000 tiny hair cells that finally turn the vibrations into electrical signals that are sent on to the brain.

Our hearing is designed to be more sensitive to high sounds than to lower sounds. If we had just a little more sensitivity to lower-pitched sounds, we would continuously be distracted by the internal sounds of our body, including the blood rushing through our arteries. In fact, to help prevent this distraction from our own body's operating sounds, there are no blood vessels at all in that part of the ear where vibrations are turned into electrical impulses. The body supports life in these tissues by constantly bathing them in dissolved nutrients.

(Reference: McCutcheon, Marc. 1989. "The Compass in Your Nose..." Los Angeles: Jeremy P. Tarcher, Inc. p. 90. Credit: Creation Moments, Inc. Foley, MN. 800-422-4253. www.creationmoments.com.)

Your Busy Liver

An adult's liver is about the size of a football and weighs about three pounds, making it the body's largest internal organ. Tucked neatly beneath the ribs, your liver performs more than 500 different tasks. It is a vital link between your heart, lungs and digestive system.

Inside the liver is a bewildering array of microscopic veins in which each drop of blood is processed. Here, blood conditions are constantly monitored to make sure everything is up to standard. If more of certain substances are needed in the blood, they are supplied. Useless chemicals are broken down into useful chemicals. Proteins are made in the liver, blood-clotting factors are corrected, hormone balances are maintained and poisons are neutralized. If substances are needed to fight an infection, they are produced and added to the blood.

The liver also stores vitamins and minerals and prepares itself to provide your body with quick energy when you need it. In addition, the liver makes bile, which is essential for digestion.

Structures like the liver have caused many evolutionists to abandon the idea that life is a result of millions of years of accidents. The liver is just too well-designed and integrated into the body to have been produced by purposelessness and mindlessness.

(Reference: "The Liver: The Body's Refinery" Discover, April 1984, p. 80. Credit: Creation Moments, Inc. Foley, MN. 800-422-4253. www. creationmoments.com.)

Your Busy Blood

Blood is one of the most miraculous of all creations. It carries oxygen and energy to our cells and carries off wastes. It is a communications pipeline, using powerful hormones to provide communication between various parts of the body.

Hemoglobin is the substance within the blood that enables it to carry oxygen to your cells. Many different kinds of creatures, including lobsters and spiders, all have some type of blood with hemoglobin in it. Some creatures even have transparent blood. Hemoglobin, all by itself, testifies to a Creator.

So many different and obviously unrelated creatures have hemoglobin that evolutionists could only account for this by saying that hemoglobin must have evolved many times in many different creatures. This explanation worked when scientists thought that hemoglobin was a relatively simple molecule.

But now we know that hemoglobin is a very complex, eight-helix twisted molecule of about a hundred atoms, all arranged in just the right way around a central atom of iron. There is a zero chance of this complex molecule happening accidentally even once, much less the many times suggested by evolutionary theory.

(Credit: Creation Moments, Inc. Foley, MN. 800-422-4253. www. creationmoments.com.)

Your Precise Blood

While most would agree that our circulatory system is vital to life, many of us don't fully appreciate the almost miraculous workings of that system. Your heart pumps about one-hundred thousand times every day. That means that your heart pumps the equivalent of 10 tons of blood every day or 80 million gallons in a lifetime.

Your circulatory system brings that blood to every cell in your body through a capillary network that is so large that the combined capillaries of only four people, stretched end to end, would reach from the Earth to the moon! But your circulatory system involves quality as well as quantity.

The chemistry of your blood is monitored and adjusted to within incredibly fine limits to keep it in precise balance second-by-second, day-by-day, year-in and year-out. For example, the acidity of your blood is adjusted constantly to within one part in a hundred million. Our circulation system cannot be simply "good enough." It has to work perfectly.

So, if mindless evolution created us, it had to make the whole system perfect the first time. There is no room for millions of years or for a poorer circulation system to be improved by mutations. Here, the facts clearly indicate that we were created in *finished-form*, not developed slowly by *trial-and-error*.

(Credit: Creation Moments, Inc. Foley, MN. 800-422-4253. www.creationmoments.com.)

~

What The Unborn Baby Tells its Mother

When talking about a baby's expected time of birth, we've all heard someone say, "The baby will come when it's ready." While that might not sound very scientific, new research shows that the statement is probably scientifically accurate. The infant, not the mother, seems to control the start of the birth process for both mother and infant.

Medical researchers are doing more research to confirm their theories. They believe that the brain of the unborn child monitors the infant's development so that it knows when the infant is ready for birth. As the time for birth approaches, a pea-sized part of the fetal brain signals the adrenal and pituitary glands that it is soon time to be born. These glands respond by producing two hormones. These hormones build up in the infant's blood. As they build up, they create changes in the mother's hormones that begin the process of giving birth.

A complete understanding of this amazing sequence may help medicine treat the problem of premature labor. More importantly, we see how exquisitely God has designed the process of giving birth. How wise of God to design the process so that when the baby is ready to be born, the mother's convenience must give way to the baby's need. Nor does the baby's birth time depend on the mother's limited understanding of the developing infant. Perhaps these realities offer some lessons to an age that places so much less value on the unborn child than God clearly does.

(Reference: Fackelmann, K. "A Fetus Tells Its Mother: It's Time for Labor." Science News, Vol. 140. P. 182. Credit: Creation Moments, Inc. Foley, MN. 800-422-4253. www.creationmoments.com.)

The Structure of Your DNA Has Not Been Left to Chance

The DNA molecule is a marvelously designed information storage system. All of the information necessary to make *you* is packed into a microscopic space. Even our most sophisticated methods of storing information cannot approach the efficiency of DNA. However, there is much more to appreciate about the structure of DNA.

Physicists used a computer to create virtual models of molecules. The molecules were depicted as beads connected to each other by tubes. Researchers were trying to find out the most efficient shape necessary to get as much molecule as possible into the smallest space. They then confined these virtual molecules within variously-shaped electronic containers.

Researchers then had the computer wiggle and slither the molecules into different shapes within each container. They were looking for the most efficient way for the molecules to fill a given space. They discovered that the most efficient shape for a molecule is the specific *helix* shape found in protein molecules.

Other research has shown that the double-helix shape that DNA uses is, likewise, the most efficient method possible to pack as much molecule into the smallest space. As an information storage system, DNA must have been clearly designed by a very wise Creator.

(Reference: Science News, 8/19/00, p. 125, "To Pack a Strand Tight, Make it a Helix." Credit: Creation Moments, Inc. Foley, MN. 800-422-4253. www. creationmoments.com.)

THE SPECTACULAR DESIGNS OF ANIMALS

Not only is the human body the result of brilliant engineering feats, so also are the varieties and intricateness of the designs seen in the animal kingdom. Consider some of these remarkable animals in the pages ahead and again ask yourself:

THE BIG QUESTION

Where did all these animals come from –
from <u>random evolution</u> or <u>intelligent design</u>?

∼

Ants Challenge Natural Selection

Charles Darwin recognized that ants challenged his theory of natural selection. He even mentioned it in his book *The Origin of The Species.* Darwin asked how the situation with the lowly ant could ever be reconciled with his theory. He never did come up with an answer, and modern evolutionists still don't have any good answers.

Darwin's problem was with the worker ants. Even though they are products of sexual reproduction, they differ greatly from their parents. They are each specialized with features their parents don't have so they can carry out their designated tasks in the nest. The problem is that these workers are sterile females, so they cannot pass on the traits that are unique from their parents.

Modern evolutionists theorize that perhaps there were some lucky mutations that took place in queen ants through their evolutionary history. However, this explanation is not very credible since the oldest fossilized ants are identical to today's ants. That means that there is no evidence of evolution in ants over a period of 70 million "evolutionary" years. Here is evidence that the ant neither evolved nor could have possibly evolved.

How many ants are there in the world? Scientists have estimated that there are about 10 million-billion ants in the world, just waiting for you to have a picnic. That's more ants than there are mammals, reptiles, birds and amphibians combined. God made so many ants because ants are important housekeepers for the earth. Ants, not earthworms, turn most of the world's soil, drain it and enrich it. Ants dispose of 90% of the corpses of small dead animals. As the world's gardeners, ants spread and plant more seeds than any other creature.

Like bees, ants have a sophisticated society, which includes workers, nurses, soldiers, hunters, farmers and even builders. Each of these different classes have specialized organs for their work. The soldier ants of one species are walking bombs. They are loaded with poison and literally explode when under attack, spraying poison on their enemies. Army ants in Central and South America march across the countryside in a line that can be nearly 50 feet long. The solid column sweeps along, flushing out even small animals as it looks for food.

The huge variety of living things around us represents more than simply variety. That variety carefully includes creatures to do every job that needs to be done to maintain life on earth. This extraordinary variety, in itself, reveals both the creativity and wisdom of the Creator.

(References: Wilson, Edward O. 1990. "Stalking the Mighty Ant." Discover, March. References: CRSnet, 2/9/00, "Evolution, Sex and the Ant." Credit: Creation Moments, Inc. Foley, MN. 800-422-4253. www. creationmoments.com.)

Ant Mathematics

Can ants count? It seems so! When scout ants find an item of food, they take it back to the nest. If the food item is especially good but too big to carry, the scout will return to the nest to get help. Scientists have discovered that ants apparently size up the task ahead before getting help so they can return with enough help, but not too much.

One scientist cut a dead grasshopper into three pieces. The second piece was twice the size of the first, and the third was twice the size of the second. He then left the pieces in different locations where ants were sure to find them. He watched as each piece was discovered by a scout, inspected, and each scout returned to the nest for help.

When the scout returned with help, the scientist counted the number of ants working at each piece of the grasshopper. The smallest piece had 28 ants working on it. The piece that was twice its size had 44 ants working on it. And how many ants do you think were working on the piece that was twice the size of the second piece? If you doubled that 44 to 88, you would be within one of being right – there were 89 ants working to return it to the nest!

We can't help but conclude that mathematical ability is part of the ants' amazing ability to plan and carry out a task.

(Credit: Creation Moments, Inc. Foley, MN. 800-422-4253. www. creationmoments.com.)

Astonishing Bee Engineers

The amazing structure of the honeycomb has fascinated scientists for thousands of years. In the third century, the astronomer and geometer, Pappus of Alexandria, became the first to offer an explanation for why the honeycomb has a hexagonal shape. Pappus explained that only three shapes could serve as candidates for a honeycomb cell – the triangle, the square, and the hexagon. Any other shape would leave wasteful open spaces between each cell. Pappus noted that the hexagon holds more honey in the same space than either a square or a triangle. It also takes less wax to build, and the shared sides of the hexagonal cells cut wax usage even further.

But it was not until the development of modern calculus that scientists could fully appreciate the shape of the caps at the end of the honeycomb cells. Each cell is capped with a pyramid composed three rhombuses. Complex mathematics shows that this shape too, requires the least amount of wax for construction and that it allows honeycomb cells to be butted up against each other without wasting space.

Modern scientists who accept evolution talk about the design of the honeycomb as a great accomplishment by bees. But the more sensible conclusion is obvious. The twelve-sided prism, that is, the six sides plus two ends of the honeycomb, is magnificent testimony to the mathematical wisdom of the Creator Himself.

(Credit: Creation Moments, Inc. Foley, MN. 800-422-4253. www. creationmoments.com.)

~

Birds Designed for Flight

Many textbooks tell young people today that birds are modified reptiles. Suppose, they say, that millions of years ago the scales on some reptiles began to fray along the edges. In time, they say, the frayed scales turned into feathers and birds were born.

The elegance and beauty of the feather make this story hard to believe. Can sticking a feather on a lizard produce a peacock? The bird's feather is only a small part of the complete flying system of the bird. Even with very careful planning and redesigning, a reptile doesn't have what it takes.

A bird needs massive breast muscles for flight. In some birds, 30% of the body weight of the bird is breast muscle. By comparison, in humans breast muscles are only about 1% of body weight. A bird also needs an extremely high metabolism and blood pressure to deliver the energy those muscles need for flight. Birds have a higher metabolism than any other creature; they also have the necessary high blood pressure.

Finally, as is well known, birds need light skeletons. The man-o'-war has a wingspan of seven feet. But its entire skeleton weighs only a few ounces – less than its feathers! Even the most-clever rebuilding of a reptile cannot produce a bird. In fact, birds have very little in common with reptiles. The entire being of the bird, from body to brain, has been specially designed for flight.

(Reference: Vandeman, George. 1991. "The Miracle of Flight. Signs of the Times," May. p. 25. Credit: Creation Moments, Inc. Foley, MN. 800-422-4253. www.creationmoments.com.)

Seals – Deep Diving Wonders

The water pressure around a human diver increases as he goes into deeper water. As the pressure increases, his blood is able to hold more dissolved oxygen. Our blood also absorbs the nitrogen in the air around us. If a diver were to move toward the surface too quickly, the nitrogen would start to bubble out of his blood. These bubbles can block the flow of blood to muscles, organs, and even the brain, leading to death. This painful condition is called the bends.

Scientists have wondered why seals don't get the bends. Weddell seals dive to far greater depths than human divers would consider using even the best equipment. The fact that they are breaking every diving rule in the book means that almost every dive should lead to a fatal case of the bends.

To find the answer to this mystery, scientists outfitted four seals with scientific backpacks. These allowed scientists to record the seals' heart rates, sample blood, and record the depths of their dives. The deeper the seals went, the more nitrogen accumulated in their blood. Just before the nitrogen reached a dangerous point it leveled off. Scientists say that the tiny sacks in the lungs that absorb oxygen and nitrogen shut down. Then the seals' heart, liver, and blubber begin to absorb the nitrogen from the blood. The air exchange sacks in the lungs reactivate as the seal ascends to the surface.

Surely the amazing biology that allows the seal to make his living deep in the ocean could only have been designed by a Creator.

(Reference: "Why Seals Don't get the Bends." Discover, October. 1985. P. 10&12. Credit: Creation Moments, Inc. Foley, MN. 800-422-4253. www. creationmoments.com.)

∾

The Efficient Fish

Fish are typical of the elegant designs found in the creation. Both function and form are united with the mechanical and biological needs of the fish to produce intelligent, beautiful, yet simple solutions for their needs. Fish are designed so that water, containing oxygen, is taken in through the mouth and expelled from behind the gill flaps. At both these points the inward and outward pressures on the fish are the greatest. The result is the most efficient system possible – many fish need expend no extra energy to breathe when they swim.

Fish tend to have their eyes located at the point on their bodies where water pressure while swimming is zero. This is important since the curvature of the cornea of the eye determines the focus of the fish's vision. It could be disastrous to the fish if its vision changed as swimming speeds varied.

The fish's heart is located over one of the points in its body where the outward pressure is greatest. While the heart muscle works by contraction, the outward pressure on the surrounding surface of the fish allows easy re-expansion of the heart for the next beat. If fish had gears and pulleys instead of their much more sophisticated biological machinery, no one would deny that it was carefully engineered by an outside intelligence

(Credit: Creation Moments, Inc. Foley, MN. 800-422-4253. www. creationmoments.com.)

The Magnificent Cheetah

Of the 41 species of cats around the world, the cheetah is one of the most unusual. Unlike other great cats, the cheetah cannot roar, but it can purr like a house cat or emit high-pitched chirps. Beautiful in form, the cheetah is the world's fastest land animal, reaching a speed of 40 miles-per-hour from a standstill within two seconds – and temporary speeds of 70 miles-per-hour.

The cheetah's claws are more like a dog's than a cat's because it is unable to retract them. It has an unusually powerful heart, an over-sized liver, extra-large and strong arteries and extra-large nostrils for taking in great quantities of air. The cheetah also has hip and shoulder girdles that swivel on its spine. As it runs, the cheetah's spine curves up and down as its legs bunch and then extend. When moving at high speed, the cheetah may only touch the ground once every 23 feet.

While most scientists believe that the cheetah evolved, the very oldest cheetah fossils show us an animal that is just about like the cheetahs we know today. This complete lack of evidence for evolution, plus the intelligent specialized features of the cheetah, lead us to the conclusion that the cheetah is the result of wise design.

(Credit: Creation Moments, Inc. Foley, MN. 800-422-4253. www. creationmoments.com.)

∽

Air-Cooled Elephants

All warm-blooded animals generate heat as they digest their food. Large animals generate a lot of heat and, like a car engine, there has to be a cooling device, otherwise the heat would kill the animal – especially in a hot climate. The elephant is a perfect example of this. It is large and lives in a hot climate.

If you were the engineer-designer of the elephant, how would you overcome the heat problem? You could provide the elephant with a lower metabolic rate so that less heat is produced in the first place. But here there is a limit. No, the solution was to equip the elephant with huge ears weighing about 100 pounds each and filled with small blood vessels.

By changing how close its ears are held to its body, or flapped, the elephant can control how much the blood in its ears is cooled before it is returned to the rest of its body. By the time our Creator finished the work of creation, He had solved millions of engineering problems just like this one. And these are often the same problems that today's engineers must solve. Is it any wonder, then, that many engineers are also Creationists?

(Reference: "How Fluids Lift Weights." Science Digest, May 1983. P. 92. Credit: Creation Moments, Inc. Foley, MN. 800-422-4253. www. creationmoments.com.)

No Worms With Kneecaps

New fossil discoveries in China are being greeted by evolutionists as among the most spectacular of the century. The fossils, say evolutionists, represent some of the earliest multi-celled creatures. Evolutionists are publicizing these fossils as evidence for evolution. However, it's not difficult to see how these fossils support creation rather than evolution.

Evolutionists admit the fossils show that the first multi-celled creatures have appeared suddenly. This confirms creationist claims that life appeared suddenly and without evolutionary ancestors. These "first" multi-celled creatures were complex and complete. There were no fossils with partially developed eyes or other organs. They include trilobites that had some of the most complex eye structures of any creature that ever lived.

There are also fossils of shrimp-like creatures and creatures with hard shells. Some of the animals were over two feet long. Now that's a big jump when you remember that the previous evolutionary step was a single-celled algae. There are no evolutionary missing links - there are no worms with kneecaps.

The evolutionists said that the creatures all belong to groups that still exist. This means that science has now documented the history of some kinds of creatures alive today right back to the very first record of multi-celled life. In other words, these creatures never evolved.

(Reference: Wilford, John Noble. "Fast Evolutionary Jump Led to Complex Life, Study Says." Star Tribune, Wednesday, April 24, 1991. P. 4A. Credit: Creation Moments, Inc. Foley, MN. 800-422-4253. www. creationmoments.com.)

Otters – An Important Extra Bone

How is it that a 50-pound sea otter can rip an abalone – its favorite food – off a rock, while a 150-pound human has a hard time cracking a simple oyster shell? The answer calls into question those evolutionary charts that compare the hand and paw bones of various mammals, including humans. Similarity of structure doesn't mean that two structures are related. For example, the octopus eye and the human eye are almost identical, but no one has suggested that we evolved from the octopus. And sometimes body structures are not as similar as evolutionists think.

Scientists have now discovered why otters can easily handle shellfish that pose problems for a human three times the otter's size. The sea otter's wrist has an extra bone in just the right place to give the otter an amazing amount of leverage in grasping shellfish.

The bone was discovered quite by accident. Scientists weren't looking for it, since their evolutionary beliefs didn't lead them to expect the extra bone. The sea otter is an example of how God has specially designed each creature. Sure, there are many similarities between creatures, since no good engineer will waste his time trying to reinvent the wheel.

(Reference: "Why Abalones Don't Find Otters Cute." Discover, Apr. 1988. p. 10. Credit: Creation Moments, Inc. Foley, MN. 800-422-4253. www. creationmoments.com.)

A Surprise Platypus

Europe was introduced to Australia's duckbill platypus in 1798. Because of the difficulties of travel in those days, scientists didn't send a live platypus from Australia to the British Museum in London. They sent only a platypus skin. Scientists in London looked at the duck bill, the beaver tail and the webbed feet of this egg laying mammal and immediately denounced the creature as a hoax! Two hundred years later the duckbill platypus continues to amaze scientists.

Not long ago, researchers discovered a surprising new ability the platypus uses to find food. It seems that the nerves in the platypus's skin, which relay the sense of touch, are also able to sense electricity. Every time we or any living creature uses a muscle, a tiny electric current is generated.

When the shrimp that the platypus eats flick their tails, they generate about 200-millionths to 1,000-millionths of a volt of electricity. That small amount of voltage is enough to enable the platypus to sense and locate lunch.

Modern biological research has also shown another mystery about the platypus. At least it's a mystery for evolutionists. While the platypus is classified as a mammal, it is genetically as different from all other mammals as mammals are from birds. Nor is the platypus genetically like the bird. This leaves the platypus with absolutely no evolutionary history, almost as if it had simply popped into existence.

(Reference: Horton, Elizabeth. 1986. "The Electric Cool Platypus." Science Digest, June. p. 21. Credit: Creation Moments, Inc. Foley, MN. 800-422-4253. www.creationmoments.com.)

Superglue Mussels

Try using superglue in wet conditions and you'll find that the glue is not so super. Even epoxy doesn't work well in water. Scientists finally decided to turn to the lowly mussel to learn how to make better glues for use in wet environments.

Mussels manufacture their glue under water, yet the glue can withstand the force of thousand pounds per square inch. Mussel glue will even stick to Teflon. The mussel begins by making the glue in two parts, each part of the glue is made by a separate gland. One gland produces proteins that are like resin. The other gland makes the hardeners.

When these are mixed, they harden into a strand in only a couple of minutes. The mussel will make many of these strands as it fastens itself to a rock. As more of these strands are made, they begin to cross-link with one another, greatly adding to the strength of the bond.

The mussel makes between five and ten different kinds of protein strands, carefully limiting the cross-linking to produce the greatest strength. Without this mix of strands, the bond would be brittle and easily break.

(Reference: Discover, 2/03, pp. 22-23, Alan Burdick, "Cement on the Half Shell." Credit: Creation Moments, Inc. Foley, MN. 800-422-4253. www. creationmoments.com.)

PLANTS, EARTH AND OTHER MARVELS
OF THE NATURAL WORLD

Of the many and extensive varieties of plant life, from the colorful and delicate petals of flowers to towering trees, consider the following examples of design and engineering complexities and ask yourself:

THE BIG QUESTION

Where did all these animals come from –
from <u>random evolution</u> or <u>intelligent design</u>?

~

The Miracle of Photosynthesis

All green plants, some algae, and even some bacteria are able to make food out of nothing more than air, water, light and a few minerals. The process is called photosynthesis, and without it we would run out of food to eat as well as oxygen to breathe.

The green plant takes in the carbon dioxide that we and animals exhale as waste and some water and, through photosynthesis, produces oxygen and a carbohydrate, and returns three-fourths of the water it originally took in for future use. Chemically what happens is that the carbon atom in carbon dioxide is removed from the oxygen and added to one water molecule, creating a carbohydrate, which is useful to us as food. While this all sounds simple, a more detailed summary of what happens each step of the way would fill a whole page with fine print.

Evolutionists marvel at the great good luck involved in the fact that plants make useful things for us out of our waste products. We have grown in our appreciation of photosynthesis since scientists have tried to mimic the process in order to build a new kind of solar cell for use in space. While the plant converts nearly 100% of the light it receives into energy, our best human efforts have reached only 8% efficiency.

So creationists ask, if the best human minds have produced only an 8% efficiency after years of work, how could no mind at all come up with nearly 100% efficiency – no matter how much time was involved?

(Credit: Creation Moments, Inc. Foley, MN. 800-422-4253. www. creationmoments.com.)

The Venus Flytrap

The exotic Venus flytrap is an insect-catching plant and a wonder of creative engineering. For example, these meat-eating plants usually live in mineral-poor soils, but by catching their own lunch, they actually provide their own fertilizer.

Up until now, scientists have not fully understood what causes the trap of the flytrap to close so rapidly on its victim. They knew that when an insect trips the little trigger hairs inside the trap, those hairs send an electric impulse to the cells on the outside of the trap.

Now it has been learned that this impulse almost instantly causes the outer cells of the trap to secrete acid. The acid breaks down the cell walls and they expand at high speed, causing the trap to close. The more the insect fights the trap, the more tightly it closes. Six to twelve hours later, after lunch is digested, the trap receives chemical signals from inside – the carnivorous plant's version of a burp – and opens to await the next meal.

(Reference: David Dreier, "Venus's Flytrap Case Closed," OMNI. Vol.5-10, July 1983, p.42. Credit: Creation Moments, Inc. Foley, MN. 800-422-4253. www.creationmoments.com.)

Scientists Recognize That We Have a Designer Earth

Evolutionists often claim that evolution can produce living things that look as if they are designed. This is their way of answering arguments that things that look designed don't need a Designer. However, in January of 2000, a paleontologist and an astronomer teamed up to publish a book that says that conditions on earth are so unique, that there is probably no other life like us in the universe. (We do need to keep in mind, however, that they write from an evolutionary perspective.)

Scientists Peter D. Ward and Donald C. Brownlee concluded that conditions could exist elsewhere in the universe that could support microbes. However, there may be nowhere else in the universe that life above that level might exist.

First, we can rule out life near the centers of galaxies because lethal levels of several types of radiation exist there. The scientists conclude that the earth is perfectly placed for intelligent life. Our moon is just the right size to control our climate, tilt and tides. The planet Jupiter acts as a giant magnetic shield to protect us from meteors and asteroids, and it, like the earth, has an orbit that does not threaten other planets. Our sun is a heavy-element star with a rare elliptical orbit. And earth's atmosphere has a carefully balanced mix of elements with just enough carbon to support life.

Biological evolution cannot explain these delicate and unusual designs, but the designs can easily be accounted for if they came from the hand of an all-wise, all-knowing Creator.

(Reference: World, 2/19/00, p. 8 "Custom-Made Earth: OK, But Who Made It?" Credit: Creation Moments, Inc. Foley, MN. 800-422-4253. www. creationmoments.com.)

The Moon Puzzle

Our moon is moving away from the earth at the rate of four centimeters per year. That might not seem like much. However, that rate of movement away from the earth presents problems for those who believe the earth and moon have been around for 4.5 billion years.

At the rate the moon is receding, it would have been so close to earth only 1.5 to 2 billion years ago that tidal friction would have melted earth's surface rocks. Mathematical fiddlings help a little, but not enough. By mathematically increasing the rate of earth's spin over supposed "billions" of years and figuring in a factor for assumed different tidal rates, one can inch the earth/moon relationship back to about 4 billion years.

Evolutionary scientists believe the problem can be solved to keep their 4.5 billion years of evolutionary history intact. But, they admit that the assumptions being tried need investigation, since the matter is far from solved. In other words, evolutionists admit they have a problem making the earth/moon relationship fit into their long-age history.

The mathematical models rule out the theory that the moon was formed billions of years ago from the same dust cloud that supposedly formed the earth. Also ruled out is the theory that the moon was captured by the earth's gravity. Only one explanation seems to satisfy the data that the moon was formed relatively recently, in cosmic time, orbiting the earth from the time of its creation.

(Reference: Dye, Brad. 1988. "The Moon Revisited." Creation Science Dialogue, Spring. p. 4. Kerr, Richard A. "Where was the Moon Eons Ago?" Science, v.221. Credit: Creation Moments, Inc. Foley, MN. 800-422-4253. www.creationmoments.com.)

~

How Do Plants Know It's Fall?

In the fall, the plants and trees of much of the Earth's northern hemisphere prepare for winter. Flowers lose their petals and the leaves fall from trees as the air begins to become cooler. But there is far more to this change in the plants than simply a response to lower temperatures.

Many, if not most, plants measure the relative length of darkness to light in a 24-hour day to tell them to prepare for winter. In a similar way, they measure the relative length of light to darkness to tell them to prepare for spring. Other living plants seem to integrate time and temperature to prepare for the seasonal change. Apple buds for example, need 1,000 to 1,400 hours of near-freezing temperatures before they even think about spring.

When winter finally arrives, we may wake up to what is a seemingly sterile world. It is then hard to believe that the plants and trees are out there sleeping while measuring time and temperature and be awake again in the spring.

(Credit: Creation Moments, Inc. Foley, MN. 800-422-4253. www. creationmoments.com.)

Hibernation: Not Simply Sleep

True hibernation is much more than just a deep, prolonged sleep. In fact, many animals that are popularly believed to hibernate, like bears, are not true hibernators. A wintering bear's body temperature seldom falls below 86 degrees, so bears are easily awakened. True hibernators actually need several hours to awaken.

The ground squirrel is a true hibernator. Its hibernation pattern is triggered by an internal clock that causes hormone changes. These changes not only lower the squirrel's temperature, metabolic breathing and heart rate, but also changes the way its nervous system and cell membranes operate. If the squirrel's nervous system and cell membrane operation were not modified for hibernation, the other changes would kill it.

Once the squirrel is in true hibernation, its body temperature drops to about 35° F, its heart rate drops from 350 beats per minute to about three beats per minute, and it will breathe only once every several minutes.

The vast and complex internal changes that must take place in hibernation, affecting the function of every cell, show us that hibernation is an ability that was built into many creatures. If all these changes were due to genetic accidents, there would be no hibernating animals today. They would all have died trying to find the right combination of internal changes to allow hibernation to occur.

(Reference: Fleming, Carol B. 1984. "How do Animals Hibernate?" Science 84, p. 28. Credit: Creation Moments, Inc. Foley, MN. 800-422-4253. www. creationmoments.com.)

New Recipe for Primordial Soup

One of the most difficult problems for those who believe in evolution is to explain the origin of life without reference to God.

Science is now aware that even the elementary living cell is, in reality, extremely complex and it has become increasingly difficult to explain this without a Creator God. The most popular explanation says that the early Earth had an atmosphere of ammonia, methane, hydrogen, and water. Oxygen had to be absent because it pretty much ruins the chemical reactions that are needed to form even simple biological molecules. It was also important that this early Earth be protected from ultraviolet radiation, which also ruins the chemistry.

But this picture of the early Earth has been shown to be totally inaccurate. For one thing, geologists can see from the oldest rocks that the earth has always had plenty of oxygen in its atmosphere. Then astronomers have pointed out that a younger sun would be turning out 10,000 times more ultraviolet radiation than it does now.

Modern science is finding out that what the Bible says is true. Scientists just won't admit it. The Earth has always had oxygen, since life-needing oxygen has been around from the first that the Earth existed. And life can only be explained as the work of the Source and Author of life, our Creator God.

(Reference: Patrick Huyghe. "New Recipe for Cosmic Soup." Science Digest, May 1983. P. 42-44. Credit: Creation Moments, Inc. Foley, MN. 800-422-4253. www.creationmoments.com.)

The Ultimate Engineering

Did you ever wonder why, while there are lots of round and cylindrical living things, there are almost no square plants or animals? Why isn't there any animal that has a skeleton made out of metal? And while there are so many ways for living things to move about, why do almost none of them have wheels?

We make lots of square things, use metal frames in the things we build, and use wheels on lots of moving things. But these features don't offer good solutions to the problems most living things have to deal with in life. Wheels are useless for going through the jungle, climbing trees, flying or burrowing. In engineering language, all living things show a high degree of design sophistication.

For example, the skeletons of all mammals have a ratio of 30% shock-absorbing collagen to 70% calcium phosphate for strength. This ratio provides the very best balance for holding up a mammal's weight during locomotion. Engineers also know that in order to get the best flow of a liquid – such as blood – a pipe's radius squared is to be equal to the sum of the radius' squared of the branches. And this is exactly the relationship found in all living and fossilized creatures, from sponges to humans!

The impressive engineering found in all living things – and even in the oldest fossils – offers elegant testimony to the Creator's wisdom and power! None of us should be shy about recognizing Him when we are with others.

(Reference: Wickelgren, Ingrid. 1989. "The Mechanics of Natural Success." Science News, v. 135, June 17. p. 376. Credit: Creation Moments, Inc. Foley, MN. 800-422-4253. www.creationmoments.com.)

~

Evolutionist Says Darwinism is in Trouble!

Several years ago, Dr. Francis Hitching gave us his book *The Neck of the Giraffe,* and in this he writes that Darwinism is in a lot of trouble. He laments, and I'm using his own words, that evolution "has not, contrary to general belief, and despite very great efforts, been proven."

Hitching points out that fossils do not show any history of evolutionary development. He says, creatures "come into the fossil record seemingly from nowhere – mysteriously, suddenly, fully formed and in a most un-Darwinian way."

He admits that the systematic gaps in the fossil record will never be filled with evolutionary ancestors and that it can no longer be claimed that someday scientists will find the missing creatures. He confesses that in the history of life, plants and animals must be treated as though they came into existence fully formed – in the forms we know today.

Hitching also notes other problems and complains that science has no idea how the genetic code could have formed without a Creator, while mutations cannot explain the supposed changes of evolution.

Hitching insists that there can be no debate that evolution actually took place. After all, he says, we are here and that is proof enough. But his statements are an honest and bottom-line admission that the theory of evolution has no support from the scientific facts.

(Reference: Francis Hitching. "Was Darwin Wrong?" Life, April, 1982. Credit: Creation Moments, Inc. Foley, MN. 800-422-4253. www. creationmoments.com.)

Fossil Inventory: Surprises for Some

Everyone will remember those school textbook diagrams showing the ever-upward progression of living organisms, including man. We recall the horse series found in textbooks and the museum displays showing the evolution of the horse: the first stage as a small mammal, and after several transitions, the modern horse. They claim that fossils in the rock layers show a progression from simple life in the lowest layers to the most complex life at the top accompanies these diagrams.

Recently, the journal, Science, reported that paleobiologists who study these fossils reevaluated all the fossil-bearing rocks that have been found in the last 180 years. What was their reaction to the meaning of the fossil record after their new inventory? "We may have been misled for twenty years," said one scientist. Another commented, "For the first time, a large group of people is saying paleobiology has been making a mistake."

Why are they reacting this way? They have had to conclude, on the basis of the fossil evidence, that there never was an ever-upward progression of complexity of life forms as they had expected. The species that are represented in the fossil record show no evidence of the classic evolutionary development traditionally found in school textbooks.

In short, the fossil record supports the Biblical claim that all the kinds of animals appeared about the same time.

(Reference: Creation, 9-11: 2001, p. 7, "Fossil Re-Count Limits Diversity." Credit: Creation Moments, Inc. Foley, MN. 800-422-4253. www. creationmoments.com.)

MASTER DESIGNS OF THE WORLD AND UNIVERSE

DR. HUGH ROSS

The following excerpt is from Dr. Hugh Ross, a distinguished astrophysicist and one of the world's foremost authorities on cosmology and the scientific evidences for Intelligent Design.
(From the book, <u>The Creator and the Cosmos</u>, by Dr. Hugh Ross, <u>NavPress Publishers, 1993</u>)

When I was eight, I started saving to buy a telescope. It took several years, but finally I pulled together enough coins to purchase the optics. With my father's help, I designed and built a mount and, at last, peered through the telescope to the heavens above.

I was stunned. I had never seen anything so beautiful, so awesome. The spectacle was too good not to share. I carried my instrument from the back yard to the front so I could invite my neighbors to join me. But no invitation was necessary. No sooner had I planted my telescope on the sidewalk than an enthusiastic crowd formed, a crowd that stayed late into the night.

That evening I began to realize many people, maybe all people, are fascinated with the starry hosts. I once thought that the sheer immensity of the heavens was responsible for that fascination. That's part of it, but there's more. There's the mystery of what's really out there, what those specks of light may be, the mystery of how they all got there and of what lies above and beyond. Gazing at the night sky seems to raise profound questions not only about the universe but also about ourselves.

The Universe and You

Cosmology is the study of the universe as a whole – its structure, origin, and development. It's not a subject just for ivory tower academics. Cosmology is for everyone.

In the words of historian, economist, and college president Dr. George Roche, "It really does matter, and matter very much, how we think about the cosmos." Roche's point is that our concept of the universe shapes our worldview, our philosophy of life, and thus our daily decisions and actions.

For example, if the universe is not created or is in some manner accidental, then it has no objective meaning, and consequently, life, including human life, has no meaning. A mechanical chain of events determines everything. Morality and religion may be temporarily useful but are ultimately irrelevant. The Universe (capital U) is ultimate reality.

On the other hand, if the universe is created, then there must be reality beyond the confines of the universe. The Creator is that ultimate reality and wields authority over all else. The Creator is the source of life and establishes its meaning and purpose. The Creator's personality defines personality. The Creator's character defines morality.

Thus, to study the origin and development of the universe is, in a sense, to investigate the basis for any meaning and purpose to life. Cosmology has deep theological and philosophical ramifications.

Unfortunately, many researches refuse to acknowledge this connection. In the name of objectivity, they gather and examine data through a special pair of glasses, the "God-is-not-necessary-to-explain-anything" glasses. It's tough for them to admit that such lenses represent their theological position, their personal faith. I've also met researchers who read the universe through the "God-is-whoever-or-whatever-I-choose" glasses.

Though no one is perfectly objective, some researchers are willing to gather and integrate the data to see which theory of origins is most consistent with the facts – whatever that theory may say about the necessity and characteristics of an Originator.

MY SKEPTICAL INQUIRY

My own thinking about the meaning of life began with my wonderment about the cosmos. I was born shortly after World War II in Montreal, Canada. My father was a self-taught engineer, and my mother a nurse. Before and during my early years, my father founded and built up a successful hydraulics engineering business. The company's rapid financial growth proved too great a temptation for Dad's financial partner, who one day withdrew all the funds and vanished. With his last few dollars, my dad brought my mother, my two sisters, and me to Vancouver, British Columbia. The neighborhood in which we settled was poor but culturally diverse. Our neighbors were mostly refugees from eastern Europe and Asia – people who, like my parents, had tasted success but either lost it or left it for survival's sake.

ARE STARS HOT?

My parents say they could see in me an intense curiosity about nature from the time I started to talk. I recall one starlit evening when I was seven, walking along the sidewalk with my parents and asking them if the stars are hot. They assured me that they are very hot. When I asked them why, they suggested I go to the library. They knew I would.

My elementary school library was well stocked with books on astronomy. As I read, I was amazed to discover just how hot the stars are and what makes them burn so brightly. I found out that our galaxy (the Milky Way) contains a hundred billion suns, and that our universe holds more than a hundred billion galaxies (now estimated to exceed 1 trillion galaxies in 2019)! I was astonished by the immensity of it all. I was compelled to find out everything I could about it.

In my eighth year, I read every book on physics and astronomy I could find in our school library. The next year I began to do the same in the children's section of the Vancouver Public Library.

By that time, I knew I wanted to be an astronomer. Many of my friends also were reading incessantly and choosing career directions. We didn't think of ourselves as precocious. The nonstop rainfall in Vancouver encouraged a lot of indoor activity and provided plenty of time to think.

At age ten, I had exhausted the science resources of the children's and youth sections of the Vancouver Public Library and was granted a pass to the adult section. A few years later, I was given access to the library of the University of British Columbia. By the time I was sixteen, I was presenting public lectures on astronomy and at seventeen won the British Columbia Science Fair for my project on variable stars. Also at seventeen, I became the director of observations for the Vancouver branch of the royal Astronomical Society of Canada (an organization of primarily amateur astronomers). I felt glad to have found so early in life a pursuit I loved.

Who Did All This?

Even as a child I always felt a sense of awe concerning nature. Its beauty and harmony, combined with its staggering complexity, left me wondering who or what could be responsible for it all.

By age fifteen, I came to understand that some form of the big bang theory provided the only reasonable explanation for the universe. If the universe arose out of a single big bang, it must have had a beginning. If it had a beginning, it must have a Beginner.

From that point on, I never doubted God's existence. But, like the astronomers whose books I read, I presumed that the Beginner was distant and non-communicative. Surely, I reasoned, a God who built a universe of more than ten-billion-trillion stars would not concern Himself with events on an insignificant speck of dust we call Earth.

Ruling Out Holy Books

My high school history studies bothered me because they showed me that the peoples of the world typically take their religions seriously. Knowing that the European philosophers of the Enlightenment largely discounted religion, I first looked for insight from their works. What I discovered, however, were circular arguments, inconsistencies, contradictions, and evasions. I began to appreciate nature all the more, for it never presented me with such twists.

Just to be fair and not to build a case on second-hand resources, I determined to investigate for myself the holy books of the world's major religions. I figured if God, the Creator, was speaking through any of these books (I presumed at the time that He was not), then the communication would be noticeably distinct from what human beings write. I reasoned that if humans invented a religion, their message would contain errors and inconsistencies, but if the Creator communicated, His message would reflect His supernature. It would be consistent like His nature is. I chose history and science as good ways to test the revelations on which various religions are based.

In the first several holy books I examined, my initial hunch was confirmed. I found statements clearly at odds with established history and science. I also noted a writing style perhaps best described as esoteric, mysterious, and vague. My great frustration was having to read so much in these books to find something stated specifically enough to be tested. The sophistry and the incongruity with established facts seemed opposite to the Creator's character as suggested to me by nature.

A Word From God?

I was getting a little smug until I picked up a Bible I had received (but not read) from the Gideons at my public school. The book's distinctives struck me immediately. It was simple, direct, and specific. I was amazed with the quantity of historical and scientific references and with the detail in them.

It took me a whole evening just to investigate the first chapter. Instead of another bizarre creation myth, here was a journal-like record of the earth's initial conditions – correctly described from the standpoint of astrophysics and geophysics – followed by a summary of the sequence of changes through which Earth came to be inhabited by living things and ultimately by humans. The account was simple, elegant, and scientifically accurate. From what I understood to be the stated viewpoint of an observer on Earth's surface, both the order and the description of creation events perfectly matched the established record of nature. I was amazed.

That night, I committed myself to spend at least an hour a day going through the Bible to test the accuracy of all its statements on science, geography, and history. I expected this study to take about four weeks. Instead, there was so much to check it took me eighteen months.

At the end of eighteen months, I had to admit to myself that I had been unsuccessful in finding a single provable error or contradiction. This is not to say that there were not any passages in the Bible I did not understand or problems that I could not resolve. The problems and passages I couldn't yet understand didn't discourage me, however, for I faced the same kinds of things in the record of nature.

But, just as with the record of nature, I was impressed with how much could be understood and resolved.

I was now convinced that the Bible was supernaturally accurate and thus supernaturally inspired. Its perfection could come only from the Creator Himself. I also recognized that the Bible stood alone in revealing God and His dealings with humans from a perspective that demanded more than just the dimensions we mortals can experience (length, width, height, time). Since humans cannot visualize phenomena in dimensions they cannot experience, finding these ideas in the Bible also argued for a superhuman author.

As a final exercise, I mathematically determined that the Bible was more reliable by far than some of the laws of physics. For example, I knew from studying physics there is roughly a one in $10\text{-}80^{th}$ (that's the number one with eighty zeros following) chance of a sudden reversal of the second law of thermodynamics. But I had calculated (with the help of skeptical friends) the probability of the chance fulfillment of thirteen Bible predictions about specific people and their specific actions.

My conservative estimate showed less than one chance in $10\text{-}138^{th}$ that such predictions could come true without supernatural intervention. That meant that the Bible was $10\text{-}58^{th}$ times more reliable than the second law of thermodynamics on just this one set of predictions. I also derived a similar conclusion based on the many instances in which the Bible accurately forecasted future scientific discoveries.

Acknowledging that my life depended moment by moment on the reliability of the second law of thermodynamics, I saw that my only rational option was to trust in the Bible's Inspirer to at least the same degree as I relied on the laws of physics. I realized, too, what a self-sufficient young man I had been.

After a long evening of studying the salvation passages in the New Testament, I humbled myself before God, asking Him to forgive me of my self-exaltation and all the offenses resulting from it, and committed myself to follow His directives for my life. At 1:06 in the morning I signed my name on the back page of my Gideon Bible, stating that I had received Christ as my Lord and Savior.

New Evidences

All of the scientific and historical evidences I had collected deeply rooted my confidence in the veracity of the bible and convinced me that the Creator had indeed communicated through this holy book. I went on to become an astronomer, and my investigations into both the cosmos and the Bible have shown me a more woundrous, personal God behind nature than I could ever have imagined.

Through the years, new evidences have consistently arisen in various fields of science, making the case for Christianity even stronger. By 1986, several breakthrough discoveries uncovered proofs for the god of the Bible so convincing that together with others I formed an organization, Reasons to Believe, to communicate these new evidences to as many people as possible.

PROBLEMS WITH DARWINIAN EVOLUTION

BY DR. RON CARLSON AND ED DECKER

The Theory of Evolution, that man and woman evolved by accident from virtually nothing into complex human beings over a period of billions of years, has been a mainstay in our society and institutions of learning since it was advanced by Charles Darwin in the mid 1800's. Throughout the 1900's and 2000's, the theory has been tenaciously assumed and supported by both novice and professional alike. No self-respecting scientist or educator would show cause for any other theory of the Origin of Life. The theory reigns supreme in the corridors of our finest schools, colleges and universities. It is the "badge of wisdom" worn by the sophisticated elite at every level and within every discipline.

The world as it exists today – with its lowered morals and standards, with its warring cultures and with its ignoble aspirations and goals – owes much to the theory of evolution. This theory is the foundation for secular self-esteem, which has been the basis for today's acceptable standard of moral and ethical living. Yes, there are many today who owe the quality of their lives to the theory of evolution. If this present philosophical trend is to continue unabated and to further evolve, it then must be recognized that the theory of evolution has experienced recent challenges in the realm of science for which there must be an answer.

This article will outline several of the scientific proofs that have been advanced against the theory of evolution, the benefits to the individual who chooses to adhere to the theory of evolution and two "key factors" (the answer) that will ensure the theory's advancement and security within our society today and into the future. This article will conclude with a personal word by the author.

DEFINITION

In a nutshell, the explanation of man as seen through the eyes of the theory of evolution is that "he is an accident that has evolved from algae." The November 20, 1989, issue of *Newsweek* magazine defined the theory in an article entitled "We're All Lucky to Be Here." Here is an excerpt:

> *It is a picture we all carry around in our heads, the most powerful icon of the Age of Secular Humanism: the Line of Ascent of Man. It begins with the bacteria, just barely across the threshold of life, a tenuous scum on the primeval seas.*

> *Then, climbing the ladder of complexity... protozoa, invertebrates, the fishes and early reptiles. Followed by the first mammals, which knocked off the dinosaurs, the early primates and their hominid descendants huddled around a cave fire. Three billion years of progress directed toward the production of Man.*

> *... This view asserts that there is nothing inherent in the laws of nature that directed evolution toward the production of human beings. There is nothing predestined about our current preeminence among large terrestrial fauna; we are the product of a whole series of contingent events in the history of our planet, any one of which could have been reversed to give rise to a different outcome.*

We are, in short, like every other creature that ever walked or slithered across the earth, an accident.

Man is nothing but an animal, the progressive result of a series of haphazard and accidental events occurring over a 3.5-billion-year span involving the first single-celled algae, which itself came from non-life, and its progression into an ever-increasing life-state of complexity. The theory of evolution essentially boils down to "Nothing + Chance = Everything."

PROBLEMATIC CHALLENGES FOR DARWINIAN EVOLUTION

The problem with the theory of evolution is that it is a 19th-century philosophical view that has been subjected to facts germinating from 20th-21st-century science, which have convinced even the most devoted followers of evolution that their conviction is based on a quagmire of misinformation and spurious data. The following are some of the more cogent opposing facts:

1. SPONTANEOUS GENERATION

Spontaneous Generation is the concept that life came from nonliving matter, a cornerstone of the evolution theory that has life emanating some 3.5 billion years ago from an inorganic gumbo of nitrogen, ammonia, salts and carbon dioxide that eventually gave birth to the a single-cell alga. Science today realizes and admits that there is no answer as to where this noxious soup came from that gave birth to this initial alga.

The idea that life arose from nonliving matter was scientifically disproved by Louis Pasteur in 1860, a fact confirmed by Dr. George Wald, Professor Emeritus of Biology at Harvard University and winner of the Noble Prize in his article entitled "Life, Origin and Evolution" published in the prestigious *Scientific American*.

Dr. Wald went on to say, "One has only to contemplate the magnitude of this task to concede that the spontaneous generation of a living organism is impossible." Biogenesis, the basic axiom of biology, dictates that life only arises from life.

2. BIOCHEMICAL REPRODUCTION

Biochemical Reproduction is the ability of a cell to extract energy from its environment in order to supply energy for the reproduction of the cell and its other needs. Science has proven that every cell has what is called a "complex metabolic motor," which performs this function; yet, this motor can only be produced by life.

This is true for DNA, deoxyribonucleic acid, the genetic code, formed in a double-helix strand, which determines the hereditary characteristics of a human being. It is essential for life to exist, but it can only be produced by life.

The question then arises, "How, when no life existed, did substances come into being which are absolutely essential for life, but which can only be produced by life?" Biochemical reproduction is the antithesis of spontaneous generation.

3. LAWS OF PROBABILITY

Laws of Probability are laws which govern the possibility of a probable result. The following is an excerpt from the book entitled "Fast Facts on False Teachings" by Dr. Ron Carlson and Ed Decker:

Another problem for the evolutionists is the laws of probability, which clearly and simply demonstrate that life from nonliving matter is impossible.

As Sir Fred Hoyle, one of the world's leading astronomers and mathematicians, recently said before the British Academy of Science: "The probability of life arising by chance is the same probability as throwing a six on a dice five million consecutive times."

Try that sometime! He then went on to add, "Let's be scientifically honest. We all know that <u>the probability of life arising to greater and greater complexity and organization by chance through evolution is the same probability as having a tornado tear through a junkyard and form out the other end a Boeing 747 Jetliner!</u>"

4. SECOND LAW OF THERMODYNAMICS

The essence of the Second Law of Thermodynamics is that everything in the universe degrades from a state of organization and complexity to a state of disorganization and chaos. This scientific law proven time and time again, without exception, flies in the face of the theory of evolution, which is based on an opposing theory that life has gone from disorganization to organization, from the simple to the complex.

The second law of thermodynamics is also expressed by the word, *entropy*, which is a thermodynamic term expressing the measure of the amount of energy unavailable for useful work in a system undergoing change – or also described as a measure of the degree of disorder in a substance or a system.

Entropy always increases and available energy diminishes in a closed system, as the universe; otherwise, someone would have produced a perpetual-motion machine a long time ago. One only has to look in a mirror over a period of time or walk through a junkyard to see this law in action.

Some evolutionists have attempted to offer the sun as a "renewable energy" source and therefore as a solution to this problem, but the facts show that random energy never produces organization. In fact, the laws of physics demonstrate that it speeds up entropy. Even more, the sun, as with the entire universe, has been shown to be "running out of steam"— going from order to chaos.

The energy that comes from any motor made by man eventually tears up the system creating the energy, making it always necessary to administer repairs.

Furthermore, just having energy is not enough. There must be multiple and complex systems in place to control energy, none of which science has been able to prove came from random processes or chance.

5. Factor of Time Assumption

The Factor of Time Assumption is the assumption that, with enough time, complexity in nature can be achieved. Another way of saying this is that, *"chance* plus *time* will generate *increased complexity."* Some evolutionists, such as astronomer Carl Sagan, subscribed to this. So, they would increasingly add billions of years to their evolutionary hypotheses. But the laws of physics prove that the greater the time span, the greater the chaos and disorganization.

6. Intelligent Design

The essence of Intelligent Design is the evidence that even the most basic bacterial cell is unexplainably complex and incorporates intricate design. Evolutionists require initial simplicity in nature in order to justify the theory that, over vast periods of time, simplicity rises to complexity. In the past, they have postulated that this "simple-to-complex" formula could be seen within the human body, as very simple cells combine to make up complex organs. But the opposite is true.

Bio-chemistry has today exposed elaborate microscopic worlds so vast that a mere thimble of cultured liquid contains over four billion single-cell bacteria, each packed with circuits, assembly instructions and miniature molecular machines. Some of these small machines function as trucks, carrying supplies from one end of the cell to the other. Others capture sunlight energy and turn it into useable energy.

There are as many machines as there are functions in the human body, e.g., hearing, seeing, smelling, tasting, healing, blood clotting, respiratory functions, etc.

One such machine is the "bacterial flagellum motor," a rotary motor that can be seen when a cell is magnified 50,000 times. An electron micrograph reveals the parts and three-dimensional structure of a motor that drives a whip-tale propeller not unlike an outboard boat motor. The motor is composed of at least 40 different protein parts, such as O-rings, drive shafts, rotors, bushings, stators, etc. If any single part is removed, the motor ceases to function – "irreducible complexity." Some of these flagellum motors have been clocked at 100,000 rpm, yet they are able to "stop on a dime" and change direction instantaneously.

All such flagellum motors are alike, which argues strongly against random chance; otherwise, there would be millions and billions of differences within these motors. The concept that just the "bacterial flagellum motor," not to mention the unnumbered other machines necessary for the function of the human body, was a product of chance is totally incongruous and mind-boggling.

Science has now shown conclusively that Charles Darwin's concept of "natural selection" could not possibly produce the "bacterial flagellum motor," much less the countless numbers of other irreducibly complex machines that function within the human body.

What is even more amazing is that this microscopic "bacterial flagellum motor" has to be produced by even smaller complex machines, which in turn had to be built by even smaller ones.

All such design must come from structure and order in assembling each part, just as a house is built from the foundation up. All such assembly order is derived from programming elements (DNA) within each human body. Programming implies a Programmer. The argument for a Programmer in this process argues strongly against evolution and its dependency on chance.

Design runs from the micro to the macro throughout the universe. When one cannot deny a designer for something as simple as a watch with all its parts, one has grave difficulty denying a Designer for something as vast and complex as the universe with all its parts that flawlessly work together.

7. FOSSIL RECORD OF TRANSITIONAL REMAINS

Charles Darwin asserted that, if evolution were true, there would exist ample evidence of transitional skeletal remains – otherwise known as "missing links" – within the earth's geological strata. He predicted that man would find a gradual upward evolution of species to greater and greater complexity among the sedimentary levels of the earth.

But the record from Mr. Darwin's writings in the 1850's to today has revealed just the opposite.

Paleontologists have devoted entire careers to looking for these transitional forms, but what they have found in the fossil records are not evidences of transitions, but species fully formed with no transitional intermediates or missing links.

If evolution were true, they would have found literally millions of transitional forms from one species to the next. The following excerpt is taken from "Acts and Facts," which was authored by Dr. Duane Gist, who received his Ph.D. in Biochemistry from the University of California at Berkeley.

> The fossil record shows the sudden appearance, fully formed, of all the complex invertebrates (snails, clams, jellyfish, sponges, worms, sea urchins, brachiopods, and trilobites) without a trace of ancestors.

> The fossil record also shows the sudden appearance, fully formed, of every major kind of fish (supposedly the first

vertebrates) without a trace of ancestors. This proves beyond a reasonable doubt that evolution has not occurred.

If evolution has occurred, our museums should contain thousands of fossils of intermediate forms. However, not a trace of an ancestor or transitional form has ever been found for any of these creatures!

Now even though evolutionary stages or links between separate species have never been proven to exist, there is ample evidence that supports evolution within a species. In fact, any person is an evolutionary step from the combination of DNA from his father and mother.

BENEFITS

There are distinct benefits afforded a person who believes in the Theory of Evolution. Those most notable benefits are listed below:

1. THE ELIMINATION OF FEAR

To the educator holding to this theory means the elimination of fear – fear from not being tenured, from not receiving research grants, from not being published and from not being accepted by associate educators.

2. FREEDOM FROM MORAL RESTRAINTS

To every individual holding to this theory the benefit is the elimination of the concept of a Creator and any responsibility to such a Creator. This in turn allows a person the freedom to think and act totally independently throughout his life and in accordance with any moral and ethical code (or lack of any) he wishes to adopt. As a world-renowned anthropologist and leading proponent of evolution once declared, "I know that the theory of evolution is scientifically impossible. However, the only alternative is to believe in a divine Creator. Therefore, I choose to believe in that which I know is false because I don't want to have any moral limitations on my sexual appetites."

3. Social Acceptance

For the individual, the belief in evolution and the rejection of a Creator will also ensure his acceptance by a world that increasingly wishes to deny a Creator.

Necessary Requirements to be Able to Successfully Accept the Theory of Evolution

From every direction during this twenty-first century, the Theory of Evolution has been scientifically refuted. But still, many wish to believe and teach it as the origin of life and man.

Where can they turn and what can they do in order to ensure they achieve their end? They must utilize two key factors to successfully reach their goal.

1. Evolutionists Must Employ the Principle of *Faith*

Regardless of how much they denigrate Christians who employ faith as a rule, evolutionists must exceed those they malign by demonstrating even more *faith*. They must accept by faith the following:

- That matter and the mind-boggling energy necessary to create the universe somehow occurred all by itself randomly and out of nothing
- That life, contrary to the proven laws of biology, somehow still occurred from non-life
- That in contrast to the mathematical laws of probability, random and impersonal chance produces increasing complexity and organization
- That somehow the second law of thermodynamics doesn't apply to evolution

- That all intermediate transitional "missing links" still exist somewhere underground but someday will be discovered

2. Evolutionists Must Rule Out The Unwanted Answer First

Evolutionists must decide first that "there is no God," and then ask the question, "What is the origin of life?" As long as they can rule out this unwanted answer up front, they may then follow whatever solution they fancy, regardless of the truth. By ruling out an undesired answer to a question, the proponents of evolutionary theory are no longer associated with education but with indoctrination. Yet, they cannot allow this hypocrisy to bother them.

Concluding Remarks

It takes more "faith" to believe the Theory of Evolution than it takes to believe in the doctrine of Creation.

The two key factors that will allow a person to accept the Theory of Evolution and be the recipient of all its benefits are *faith in its scientific impossibilities* and a willingness to accept *anything but Biblical truth.*

If a person will employ these two factors, he or she will have smooth sailing in the arena of evolution, but they will surely suffer in purpose, identity and self-esteem.

GOD IS INVITING YOU TO A PERSONAL ENCOUNTER

SECTION 3

Let's recap where we are at this point in your *Encounters* experience.

In the Introduction, we talked about how there is more at work in this world than just those things we can see, touch or experience physically. To the person who says, "If I can't see it, I can't believe it," we talked about how quickly that logic fails when you stick a paper clip in a wall outlet, jump off a building or turn on a radio. Electricity, gravity and radio waves are all clear and indisputable realities, even though we can't see or touch them.

We then explored dozens of remarkable stories described as "miracles" or "supernatural" encounters. Most of these stories defy any possibility of mere coincidence or happenstance. They were *beyond*-natural, or *super*-natural. As you read them, how was your heart responding? Did you find yourself in awe or having wonderment? Were your emotions touched – any tears or thrills? At any point, especially if you are an atheist or skeptic, did you find yourself thinking, "Maybe there's more out there than I thought?"

As you read the encounter stories, it's evident that deeper unseen forces are at work. These are spiritual forces, not physical. All of those involved in these stories felt that their experiences were the work of God intervening in their lives.

But what about the person who refuses to accept this premise, who sincerely believes that there is no God and that all in this world is the result of natural processes?

That's why we walked through the next section of the book on the brilliant intelligent designs of the human body, animals and nature.

Additionally, we looked at the substantial problems with Darwinian evolution and how it takes more *faith* to believe that all of life arose randomly by accident in increasing complexity, defying known laws of science, than it does to believe in an intelligent Creator – aka God.

Just like the example of finding a perfect Rolex watch on the hiking trail, an intellectually honest person would have to admit that the Rolex was the result of intelligent design, not random evolution. Otherwise, if it were just random evolution, we'd see Rolex watches popping up all over the place!

Similarly, if Darwinian evolution is true, we'd also see numerous examples around us of animals in their current evolutionary transition stages between species groups – frogs with wings as they are transitioning to birds, fish with legs as they are transitioning to land mammals, even Neanderthal cave men as they are transitioning from monkeys to modern humans.

But we don't see any of these evolving transitions.

More troublesome for evolutionists, there is *no example* in the extensive world-wide fossil record of *any* transitional or "missing" links between species.

Though there are mutations and adaptations within the same species groups, there is *no evidence* of any evolutionary development across species – such as a fish with fins that evolves into a land mammal with legs. This enormous lack of transitional evidence is devastating to evolutionary theory. And this creates a major dilemma.

The Dilemma

How does a person who has sincerely believed in natural evolution respond to The Big Question of *Where Did We Come From*? There are only two options: *Random Evolution* or *Intelligent Design*.

An intellectually honest person would have to agree that the far-greater complexity and engineering we observe just in the human body defies random evolutionary theory. We can't see electricity, but we can observe its evidence. So it is with God – we can't see Him, but we can intelligently discern His fingerprints that are clearly revealed in the brilliant designs of man, animals and nature.

But how do we know that the Intelligent Creator is the God revealed in the Bible and not some god or other spiritual force connected to Hinduism, Buddhism, Islam or another religious philosophy?

Great question. Do your own study and explore the evidence. The evidence is substantial. In fact, the last section of this book, Section 4, presents extraordinary evidence about the historical and documentary accuracy of the Bible and the resurrection claims of Jesus, with significant evidence that the Creator is none other than the one whom the Bible calls God.

So where do we go from here?

Encounter Your Creator

The God of the universe, the intelligent Creator who designed you and formed you, He *knows* you – and He *loves* you. He knows your name and everything about you. In fact, He's crazy about you!

He's given you a mind to think and make choices. He's given you a heart to feel emotions. He's placed within you a longing that only He can fill. No money, status, possession, fame, relationship or human achievement can ever fully satisfy the deepest longings of your soul.

Look no further than the celebrity tabloids facing you at the grocery checkout lines. These celebrities have it all - money, fame, beauty, mansions, yachts, glitzy parties, exotic vacations - and yet very few seem truly happy. Scan the tabloid stories and you'll see that they're filled with sordid life dramas and relationship miseries. How is this possible for those people who "have everything?"

One of my friends from childhood is a famous Hollywood actress. She's been in a number of television shows and movies. Her face has been on dozens of magazine covers - *Cosmopolitan, Vogue, People, InStyle* etc. She's beautiful, famous and wealthy. Yet, her eyes have always looked empty. She's never seemed happy.

I'm closer to her older sister, and years ago, her older sister and I went to visit her at her house in Bel Air, California. It was an afternoon pool party with a lot of her sister's Hollywood friends. The interesting thing to me was how few of these Hollywood actors and actresses seemed to *actually be happy*. Why? They had it all!

The older sister is a Christian missionary, and I've frequently marveled at how she, *the missionary with no money or fame*, possesses so much more *peace, joy, purpose, contentment* and *true happiness* than her famous, wealthy celebrity sister. The celebrity sister "has it all," but seems *empty*. The missionary sister has little, but is *full*. Why?

As the renowned French scientist Blaise Pascal said, "There is a God-shaped vacuum in the heart of each man which cannot be satisfied by any created thing, but only by God the Creator, made known through Jesus Christ."

God wants a personal *relationship* with *you*. He wants you to know Him, so He can fill you with all of the deep longings of your heart.

There's so much He desires to give to you. He wants to fully satisfy your deepest desires to *love* and be *loved*... to *know* and be *known*... to be *cherished*... *respected*... *honored*... to have true *purpose* and *meaning*... to be *secure*... to have in full measure *peace, wholeness* and a sense of *completion* – that state of being described by Jews as *Shalom*.

In short, God wants you to have an *encounter* with Him.

Right here... Now.

Have you considered that this may be the reason you are reading this book? Has He placed this book in your hands *at this very time* for a higher purpose? Maybe He's brought you to this specific moment so you can finally encounter Him and His love for you?

God doesn't just want to bring good things to your life, He wants you to know Him and enjoy Him. Not just now, but *forever*.

God is eternal in nature, and so are you. You have a physical body plus a mind, will and emotions. However, the core of who you are is *spiritual*, and that *spiritual* core – your *true self* – is also eternal in nature. You aren't just a body with a soul, you are literally an *eternal soul* temporarily wrapped in a physical body.

When your body dies, your true self, your soul, continues living forever. It never dies. But when your present body dies, what happens then? Where will you go?

God through His own Word clearly answers that question. There are only two destinations, Heaven or Hell, and we get to choose which one we'll go to. Both are real. You're on the path right now toward one or the other. The choice is yours.

(If you're thinking that the concepts of Heaven and Hell are simply allegorical, wives' tales or old-fashioned concepts, please read the Author's Closing Comments section at the end of this book.)

God offers eternal life as a *free gift* to any who choose Him! His heart is to be loved by man, but in order for this love to be meaningful, it has to be man's choice. It can't be robotic or forced if it is to mean anything. So, when God created man, He gave us the ability to make independent choices.

A key in understanding how to enter into a personal relationship with the God who loves you is to understand some important fundamentals. First, God is Holy, pure and perfect. Second, man is not. You and I fall far short of God's standard of perfect obedience to the moral law He has established, both in our actions and in our heart motivations.

All of us have broken this moral law and, therefore, are unable to have a direct relationship with Him.

In fact, the consequence of breaking even one of His requirements of goodness is the inability to go to Heaven, which is reserved only for those who are completely righteous in God's sight.

Any failure on any point of God's perfect righteous standard is called sin. The Bible declares that all have sinned and fallen short of God's perfect standard, and that none is righteous, not even one person.

The Bible clearly states that the penalty for even committing one sin in an entire lifetime is death, meaning eternal death – spending an eternity in the most awful place imaginable called Hell.

> *"For all have sinned and fallen short of the glory of God."*
> *Romans 3:23*

> *"There is none righteous, no not one."*
> *Romans 3:10*

> *"For the wages of sin is death,*
> *but the free gift of God is eternal life in Jesus Christ our Lord."*
> *Romans 6:23*

But because of God's great love for you and for me, He's made a provision that gives each person an opportunity to not only go to Heaven, but to also enjoy an abundant life while here on earth.

That provision was through His son, Jesus Christ, the long-awaited and prophesied-about Messiah of the Scriptures.

"For God so loved the world that He gave His only son, that
whoever believes in him shall not die, but have everlasting life."
John 3:16

"God demonstrates His great love for us in that, while we were yet
sinners, Christ died for us."
Romans 5:89

Jesus Christ chose to come to earth because of His deep love for you. He was fully God and fully man at the same time, something we can't understand with our finite minds. He didn't come to force people to follow him; He came to die – for you. And because Jesus was able to live an absolutely sin-free righteous life, fulfilling more than 40 Jewish Messianic prophecies as their future Messiah, God accepted his death in place of yours.

He went one step further and designed it so that *anyone* who chose to accept Christ into their life as Lord and ruler of their life could receive salvation – which means eternal life in Heaven upon death of their physical body.

Yes, Heaven is only for those who are righteous, but as a follower of Christ, Christ's perfect and righteous life transfers to them.

When God sees a true Christian man or woman, He sees the righteous perfection of His Son in them, who already paid the death penalty, and as a result, God and man can enter into intimate relationship.

"I have come that you might have life and have life abundantly."
John 10:10

"Yet to all who did receive Him (Jesus and his death for your sins in
your place), to those who believed in His Name,
He gave the right to become children of God."
John 1:12

"Everyone who calls on the Name of the Lord will be saved."
Romans 10:13

"If you declare with your mouth, "Jesus is Lord" (meaning your
master) and believe in your heart that God raised him from the
dead, you will be saved. For it is with your heart that you believe
and are justified, and it is with your
mouth that you profess your faith and are saved."
Romans 10:9,10

"For it is by grace you have been saved through faith – and this is
not from yourselves, it is the gift of God – not by your works,
so that no one can boast."
Ephesians 2:8,9

So, the choice is yours.

It is truly THE MOST IMPORTANT DECISION of your life, for it determines not only how the rest of your life will go here on earth, but more importantly, where you will spend your eternal destination.

Heaven and Hell literally hang in the balance.

Time to Respond

If you haven't made a commitment to accept God's free gift of salvation and serve Him as the Lord of your life ("Lord" meaning your master and the one you serve), it's very simple. Just ask Him by prayer, which is nothing more than talking to God. He's always present around you and will hear whatever you say. Simply say something like the following, but make sure you *sincerely* mean what you're saying:

Dear Lord Jesus. Thank you for loving me and opening my eyes to
your truth. I know that I am a sinner, and I ask for Your
forgiveness. I believe You died for my sins in my place and rose from

the dead. I turn away from my sins and invite You to come into my heart and life. I commit to trust and follow You as my Lord and Savior. In Your Name I commit this. Amen.

Seriously, that's it. If you prayed this prayer sincerely and are committed to trusting and following Christ as your Lord, God has already responded and will begin to do marvelous things in your life.

You have become what is called "born again." Your eternal destination has just shifted from Hell to Heaven, and you are now an adopted son or daughter of the God who created you.

> *"Therefore, if anyone is in Christ, the new creation has come:*
> *The old has gone, the new is here!"*
> *2 Corinthians 5:17*

As a follower of Christ, be committed. Spend time in His Word and in prayer to grow. Find other Christian believers who are solid in their faith and connect with them. Find a solid church that esteems the scriptures of the Bible. There's much you need to learn going forward, especially if you want to experience the fullness of what God offers to you.

If you want more information on how to engage in an authentic deeper relationship with Christ, or for any similar topics, email:

EncountersBookConversation@gmail.com

A BRILLIANT ATHEIST EXAMINES THE EVIDENCE FOR CHRISTIANITY

SECTION 4

Josh McDowell, the author of this final section, is a long-time friend. He is a brilliant thinker, a former atheist and is considered one of the greatest Christian apologists of the 20th and 21st centuries. Over the past 50 years, Josh has spoken to more than ten million young people in 84 countries on more than 700 university and college campuses. He has authored or coauthored over 150 books with more than fifty million in print worldwide. His most popular books are *Evidence That Demands a Verdict, More Than a Carpenter, Don't Check Your Brains at the Door* and *The Last Christian Generation*.

The following is an intriguing story of how Josh sought to disprove the resurrection of Jesus Christ, the singular event upon which *all of Christianity* hinges. The results of his effort – involving over 1,000 hours of research to disprove the Resurrection – changed his life and literally millions of others', including mine. It's an excellent presentation of evidence arranged like a defense lawyer might present in a court of law that was convened to evaluate the evidentiary merits of Christianity. Specifically, what historical evidence is there to support the radical claim that Jesus was resurrected from death to life.

Why is the Resurrection such a significant event?

Because, if Jesus actually raised from death to life, it would validate his life's teachings, the truth of the scriptures of the Bible, his claim to be God in the flesh and his claim that the only way to attain eternal life is through him.

No other founder or leaders of world religions – Islam, Buddhism, Hinduism, etc. – ever came back to life after death. Jesus stands alone in this category, and as you'll read, the evidence for his life, death and resurrection back to life is overwhelming.

> *"I am the Way, the Truth and the Life.*
> *No one comes to the Father except through me."*
> *John 14:6*

CHRISTIANITY - HOAX OR HISTORY?

DR. JOSH MCDOWELL

HE CHANGED MY LIFE

Thomas Aquinas wrote: "There is within every soul a thirst for happiness and meaning."

I wanted to be happy. There's nothing wrong with that. I also wanted to find meaning in life. I wanted answers to the questions: Who am I? Why in the world am I here? Where am I going?

More than that, I wanted to be free. Freedom to me was not going out and doing what I wanted to do. Freedom was having the power to do what I knew I ought to do . . . but didn't have the power to do. So I started looking for answers. It seemed that almost everyone was into some sort of religion, so I did the obvious thing and took off for church.

I must have found the wrong church, though. Some may know what I mean: I felt worse inside the church than I did outside.

I've always been very practical, and when one thing doesn't work, I chuck it. So, as a young college student, I chucked religion. The only thing I had ever gotten out of religion was the change I took out of an offering to buy a milkshake! And that's about all many people ever gain from "religion."

I began to wonder if prestige was the answer. So in college, I ran for freshman class president and got elected. It was neat knowing everyone on campus, having everyone say, "Hi, Josh," making the decisions, spending the university's money and the students' money to get speakers I wanted. It was great, but it wore off like everything else I had tried.

I was like a boat out in the ocean being tossed back and forth by the waves, the circumstances. And I couldn't find anyone who could tell me how to live differently or give me the strength to do it. Then I began to notice people who seemed to be riding above the circumstances of university life. One important thing I noticed was that they seemed to possess an inner, constant source of joy – a state of mind not dependent on their surroundings. They were disgustingly happy. They had something I didn't have . . . and I wanted it.

I began purposely to spend more time with these people, and we ended up sitting around a table in the student union one afternoon. Finally, I leaned back in my chair and said, "Tell me, have you always been this way, or has something changed your lives? Why are you so different from the other students, the leaders on campus, the professors? Why?"

One student looked me straight in the eye with a little smile and said two words I never thought I'd hear as part of any solution in a university. She said, "Jesus Christ."

I said, "For God's sake, don't give me that garbage. I'm fed up with religion; I'm fed up with the church. Don't give me that garbage about religion."

She shot back, "I didn't say 'religion'; I said, 'Jesus Christ.'"

It wasn't long before these new friends challenged me intellectually to examine the claims that Jesus Christ is God's Son, that He took on human flesh, that He lived among real men and women and died on the cross for the sins of mankind, that He was buried, and that He arose three days later and could change a person's life in the twentieth century.

Finally, I accepted their challenge. I did it out of pride, to refute them. But I didn't know there were facts. I didn't know there was evidence that a person could evaluate.

As I delved into my research on Christ, I discovered that men and women down through the ages have been divided over the question, "Who is Jesus?" It didn't take long for the people who knew Jesus to realize that He was making astounding claims about Himself. Especially during the trial of Jesus – the trial that eventually led Him to the cross — I found one of the clearest references to Jesus' claims of deity.

Then the High Priest asked him, "Are you the Messiah, the Son of God?" Jesus said, "I am, and you will see me sitting at the right hand of God, and returning to earth in the clouds of heaven" (Mark 14:61-62). Jesus claimed to be God. He didn't leave any other option open. His claim must either be true or false. Jesus' question to His disciples, "Who do you think I am?" (Matthew 16:15) has several possibilities.

WAS HE A LIAR?

If, when Jesus made His claims, He knew that He was not God, then He was lying and deliberately deceiving His followers. And if He was a liar, then He was also a hypocrite because He told others to be honest, whatever the cost, while He Himself taught and lived a colossal lie.

This view of Jesus, however, doesn't coincide with what we know either of Him or of the results of His life and teachings. Whenever Jesus has been proclaimed, lives have been changed for the good, nations have been changed for the better. Thieves have been made honest, alcoholics have been cured, hateful individuals have become channels of love, unjust persons have become just.

William Lecky, one of Great Britain's most noted historians and a dedicated opponent of organized Christianity, wrote about Jesus' ministry: "The simple record of these three short years of active life has done more to regenerate and soften mankind than all the discourses of philosophers and all the exhortations of moralists."[1]

Someone who lived as Jesus lived, taught as Jesus taught, and died as Jesus died could not have been a liar. What other alternatives are there?

WAS HE A LUNATIC?

If it is inconceivable for Jesus to be a liar, then couldn't He actually have thought Himself to be God but been mistaken? After all, it's possible to be sincere and wrong.

Someone who believes he is God sounds like someone today believing himself to be Napoleon. He would be deluded and self-deceived and probably would be locked up so he wouldn't hurt himself or anyone else. Yet, in Jesus we don't observe the abnormalities and imbalances that usually go along with being deranged. His poise and composure when confronted by His enemies would certainly be amazing if He were insane.

Here is a man who spoke some of the most profound sayings ever recorded. His instructions have liberated many individuals in mental bondage.

A student at a California university told me that his psychology professor had said in class that "all he has to do is pick up the Bible and read portions of Christ's teachings to many of his patients. That's all the counseling they need."

Psychiatrist J. T. Fisher, speaking of Jesus' popular "Sermon on the Mount" (Matthew 5-7), says this: "For nearly two thousand years the

Christian world has been holding in its hands the complete answer to its restlessness and fruitless yearnings. Here ... rests the blueprint for successful human life with optimism, mental health, and contentment."[2]

WAS HE LORD?

I cannot personally conclude that Jesus was a liar or a lunatic. The only other alternative is that He is the Christ – the Son of God – as He claimed to be.

When I discuss this with many people, it's interesting how they respond. I share with them the claims Jesus made about Himself and then the material about Jesus being a liar, lunatic, or Lord. When I ask if they believe Jesus was a liar, there is usually a sharp, "No!"

Then I ask, "Do you believe He was a lunatic?" The reply is, "Of course not." Then, "Do you believe He is God?" But before I can get a breath in edgewise, there is a resounding, "Absolutely not." Yet, one has only so many choices. One of these options must be true.

The issue with these three alternatives is not *which is possible?* for it is obvious that any of the three could have been possible. But, rather, it is the question *Which is more probable?*

Who you decide Jesus Christ is must not be an idle intellectual exercise. You cannot put Him on the shelf while calling Him a great moral teacher. That is not a valid option, because if He was so great and moral, what are you going to do with His claim to be God?

If He was a liar or lunatic, then He can't qualify as a great moral teacher. And if He was a great moral teacher, then He is much more as well. He is either a liar, a lunatic, or the Lord God. You must make a choice.

"But," as the Apostle John wrote, "these are recorded so that you will believe that he is the Messiah, the Son of God, and [more important] that believing in him you will have life" (John 20:31).

Two issues became clear in my study of Christianity:

1. Is CHRIST'S RESURRECTION HISTORICALLY CREDIBLE?

This is crucial because Christ appealed to His resurrection as the proof that His claims of deity were true.

1. Is THE NEW TESTAMENT ACCOUNT OF CHRIST RELIABLE?

Most of what we know about Christ comes from the New Testament. So, is the New Testament account of Christ reliable—can it be trusted?

These are the questions that I want to address in the remainder of this section.

CHRISTIANITY HOAX OR HISTORY?

BACK FROM THE GRAVE

For centuries, many of the world's most distinguished philosophers have assaulted Christianity as being irrational, superstitious and absurd. Many have chosen simply to ignore the central issue of the Resurrection. Others have tried to explain it away through various theories. But the historical evidence just can't be discounted. Confronting the facts of the empty tomb is as convincing today as it was 2,000 years ago.

A QUESTION OF HISTORY

A student at the University of Uruguay said to me: "Professor McDowell, why can't you refute Christianity?" "For a very simple reason," I answered. "I am not able to explain away an event in history – the resurrection of Jesus Christ."

How can we explain the empty tomb? Can it possibly be accounted for by any natural cause? Here are some of the facts relevant to the Resurrection:

1. Jesus of Nazareth, a Jewish prophet who claimed to be the Christ prophesied in the Jewish Scriptures, was arrested, was judged a political criminal and was executed by Roman crucifixion.

2. Three days after His death and burial, some women who went to His tomb found the body gone.

3. In subsequent weeks, His disciples claimed that God had raised Him from the dead and that He appeared to them at various times before ascending into heaven.

4. From that foundation, Christianity spread throughout the Roman Empire and has continued to exert great influence down through the centuries. Did the Resurrection actually happen?

5. Was the tomb of Jesus really empty? Those questions raise controversy even today.

After devoting more than 1,000 hours of studying this subject, I have come to the conclusion that the resurrection of Jesus Christ is either one of the most heartless hoaxes ever foisted on the minds of human beings – or it is the most remarkable fact of history. The Resurrection issue takes the question, "Is Christianity valid?" out of the realm of philosophy and forces it to be a question of history.

Does Christianity have a historically acceptable basis? Is sufficient evidence available to warrant belief in the Resurrection?

A. Is The New Testament Reliable?

Because the New Testament provides the primary historical source for information on the Resurrection, many critics during the nineteenth century attacked the reliability of these biblical documents.

By the end of the nineteenth century, however, archaeological discoveries had confirmed the accuracy of the New Testament manuscripts; many places, events and people referred to in the New Testament turned out to be true. Discoveries of early papyri manuscripts have also helped bridge the gap between the time of Christ and existing manuscripts from a later date.

Those findings increased scholarly confidence in the reliability of the Bible. William F. Albright, who in his day was the world's foremost biblical archaeologist, said:

> "We can already say emphatically that there is no longer any solid basis for dating any book of the New Testament after about AD. 80 – two full generations before the date between 130 and 150 given by the more radical New Testament critics of today."[1]

Coinciding with the papyri discoveries, an abundance of other manuscripts came to light. (More than 24,000 copies of early New Testament manuscripts are known to be in existence today.) That fact motivated Sir Frederick Kenyon, one of the leading authorities on the reliability of ancient manuscripts, to write:

> The interval then between the dates of original composition and the earliest extant evidence becomes so small as to be in fact negligible, and the last foundation for any doubt that the Scriptures have come down to us substantially as they were written has now been removed. Both the authenticity and the general integrity of the books of the New Testament may be regarded as finally established.[2]

The historian Luke wrote of "authentic evidence" concerning the Resurrection. Sir William Ramsey, who attempted for fifteen years to undermine Luke's credentials as a historian and to refute the New Testament's reliability, finally concluded: "Luke is a historian of the first rank... This author should be placed along with the very greatest of historians."[3]

B. LIVING WITNESSES

The New Testament accounts of the Resurrection were being circulated within the lifetimes of men and women alive at the time of the event. Those people could certainly have confirmed or denied the accuracy of such accounts.

The writers of the four Gospels either had themselves been witnesses or else were relating the accounts of eyewitnesses of the actual events. In advocating their case for the gospel, a word that means "good news," the apostles appealed (even when confronting their most severe opponents) to the common knowledge concerning the facts of the Resurrection.

F. F. Bruce, professor of biblical criticism and exegesis at the University of Manchester, says concerning the value of the New Testament records as primary sources:

> "Had there been any tendency to depart from the facts in any material respect, the possible presence of hostile witnesses in the audience would have served as a further corrective."[4]

The facts and details of what Christ had said and done were presented in the very presence of antagonistic eyewitnesses of Christ who knew the events surrounding Christ's life and ministry. In that you have historically what we call today in a court of law the principle of "cross-examination" to discern truth from fabrication.

C. Background

The New Testament witnesses were fully aware of the background against which the Resurrection took place. The body of Jesus, in accordance with Jewish burial custom, was wrapped in a linen cloth. About 100 pounds of aromatic spices, mixed together to form a gummy or cement-like substance, were applied to the wrappings of cloth about the body to form an encasement weighing about 120 pounds.

After the body was placed in a solid rock tomb, the historical account points out that an extremely large stone closed the entrance of the tomb. The large stone weighed approximately one-and-a-half to two tons and was rolled (by means of levers) against the tomb's entrance.

A Roman guard unit of sixteen strictly disciplined fighting men was stationed to guard the tomb. This guard unit affixed on the tomb the Roman seal, which was meant to prevent any attempt at vandalizing the sepulcher. Anyone trying to move the stone from the tomb's entrance would have broken the seal and incurred the wrath of Roman law.

But three days later the tomb was empty. The followers of Jesus said He had risen from the dead. They reported that He appeared to them during a period of forty days, showing Himself to them by many "infallible proofs." Paul the apostle recounted that Jesus appeared to more than 500 of His followers at one time, the majority of whom were still alive and could confirm what Paul wrote. No one acquainted with the facts can accurately say that Jesus appeared to just "an insignificant few."

D. ATTEMPTED EXPLANATIONS

Christians believe that Jesus was bodily resurrected in time and space by the supernatural power of God. The difficulties of belief may be great, but the problems inherent in unbelief present even greater difficulties. Put another way, when it comes to the Resurrection, *the burden of unbelief is greater than the burden of belief.* The theories advanced to explain the Resurrection by "natural causes" are weak; they actually help to build confidence in the truth of the Resurrection.

1. THE WRONG TOMB?

A theory propounded by Kirsopp Lake assumes that the women who reported the body was missing had mistakenly gone to the wrong tomb. If so, then the disciples who went to check up on the women's statement must have also gone to the wrong tomb. We may be certain, however, that the Jewish authorities, who asked for a Roman guard to be stationed at the tomb to prevent Jesus' body from being stolen, would not have been mistaken about the location. Nor would the Roman guards, for they were there!

If the Resurrection claim was merely because of a geographical mistake, the Jewish authorities would have lost no time in producing the body from the proper tomb, thus effectively squelching for all time any rumor of resurrection.

But what did the soldiers and the Jewish authorities do? The record states that...

Some of the guards went into the city and reported to the chief priests everything that had happened. When the chief priests had met with the elders and devised a plan, they gave the soldiers a large sum of money, telling them, "You are to say, 'His disciples came during the night and stole him away while we were asleep.' If this report gets to the governor, we will satisfy him and keep you out of trouble." So the soldiers took the money and did as they were instructed. And this story has been widely circulated among the Jews to this very day. (Matthew 28:11-15, NIV)

2. The Body Stolen?

Consider the theory that the body was stolen by the disciples while the guards slept. As the Scriptures note, this is the very oldest attempted explanation.

However, the depression and cowardice of the disciples provide a hard-hitting argument against their suddenly becoming so brave and daring as to face a detachment of soldiers at the tomb and steal the body. They were in no mood to attempt something like that.

J. N. D. Anderson has been dean of the faculty of law at the University of London and director of its Institute of Advanced Legal Studies. Commenting on the proposition that the disciples stole Christ's body, he says:

> "This would run totally contrary to all we know of them: their ethical teaching, the quality of their lives, their steadfastness in suffering and persecution. Nor would it begin to explain their dramatic transformation from dejected and dispirited escapists into witnesses whom no opposition could muzzle."[5]

An alternative theory that the Jewish or Roman authorities moved Christ's body is no more reasonable an explanation for the empty tomb than theft by the disciples. If the authorities had the body in their possession or knew where it was, why, when the disciples were preaching the Resurrection in Jerusalem, didn't they explain: "Wait! We moved the body. He didn't rise from the grave"?

And if such a rebuttal failed, why didn't they explain exactly where Jesus' body lay? If this failed, why didn't they recover the corpse, put it on a cart, and wheel it through the center of Jerusalem? Such an action would have destroyed Christianity – not in the cradle but in the womb!

Dr. John Warwick Montgomery, an attorney and dean of the Simon Greenleaf School of Law, further explains,

> "It passes the bounds of credibility that the early Christians could have manufactured such a tale and then preached it among those who might easily have refuted it simply by producing the body of Jesus."[6]

3. HALLUCINATIONS?

One of the most desperate appeals to explain away the Resurrection is the appeal to hallucinations. In no way can one say that Jesus' appearances were stereotyped or that His followers were hallucinating what happened to them according to some trumped-up formula intended to convince people of what was actually not so.

The American Psychiatric Association's official glossary defines a "hallucination as a false sensory perception in the absence of an actual external stimulus."

Hallucinations are linked to an individual's subconscious and to his or her particular past experiences, making it very unlikely that even two people could have the same hallucination at the same time. Christ appeared to many people, and descriptions of the appearance involve great detail, like those which psychologists regard as determined by reality.

Christ also ate with those to whom He appeared. And He not only exhibited His wounds, but He also encouraged a closer inspection. An illusion does not sit down and have dinner with you, and it cannot be scrutinized by various individuals at will.

A hallucination is a very private event – a purely subjective experience void of any external reference or object. If two people cannot initiate or sustain the same vision without any external object or reference, how could more than five hundred do so at one time? This is not only contrary to this principle of hallucinations but also strongly mitigates against it. The many claimed hallucinations would be a far greater miracle than the miracle of resurrection. This is what makes the view that Christ's appearances were hallucinations so ludicrous.

4. DID JESUS SWOON?

Another theory was popularized by Venturini several centuries ago and is often quoted today. This is the swoon theory, which says that Jesus didn't die; he merely fainted from exhaustion and loss of blood. Everyone thought He was dead, but later He resuscitated, and the disciples thought it to be a resurrection.

Skeptic David Friedrich Strauss – certainly no believer in the Resurrection – gave the deathblow to any thought that Jesus revived from a swoon:

> "It is impossible that a being who had stolen half-dead out of the sepulcher, who crept about weak and ill, wanting medical treatment, who required bandaging, strengthening and indulgence, and who still at last yielded to His sufferings, could have given to the disciples the impression that He was a Conqueror over death and the grave, the Prince of Life, an impression which lay at the bottom of their future ministry. Such a resuscitation could only have weakened the impression which He had made upon them in life and in death, at the most could only have given it an elegiac voice but could by no possibility have changed their sorrow into enthusiasm, have elevated their reverence into worship."[7]

A PHYSIOLOGIST LOOKS AT THE CRUCIFIXION

Samuel Houghton, M.D., the great physiologist from the University of Dublin, relates his view on the physical cause of Christ's death:

> "When the soldier pierced with his spear the side of Christ, He was already dead; and the flow of blood and water that followed was either a natural phenomenon explicable by natural causes or it was a miracle . . . .

> Repeated observations and experiments made upon men and animals have led me to the following results. When the left side is freely pierced after death by a large knife, comparable in size with a Roman spear, three distinct cases may be noted:

> First. No flow of any kind follows the wound, except a slight trickling of blood. Second. A copious flow of blood only follows the wound. Third. A flow of water only, succeeded by a few drops of blood, follows the wound.

Of these three cases, the first is that which usually occurs; the second is found in cases of death by drowning and by strychnia, and may be demonstrated by destroying an animal with that poison, and it can be proved to be the natural case of a crucified person; and the third is found in cases of death from pleurisy, pericarditis, and rupture of the heart.

With the foregoing cases most anatomists who have devoted their attention to this subject are familiar; but the two following cases, although readily explicable on physiological principles, are not recorded in the books (except by St. John). Nor have I been fortunate enough to meet with them.

Fourth. A copious flow of water, succeeded by a copious flow of blood, follows the wound. Fifth. A copious flow of blood, succeeded by a copious flow of water, follows the wound.

Death by crucifixion causes a condition of blood in the lungs similar to that produced by drowning and strychnia; the fourth case would occur in a crucified person who had previously to crucifixion suffered from pleuritic effusion; and the fifth case would occur in a crucified person, who had died upon the cross from rupture of the heart.

The history of the days preceding our Lord's crucifixion effectually excludes the supposition of pleurisy, which is also out of the question if blood first and water afterwards followed the wound. There remains, therefore, no supposition possible to explain the recorded phenomenon except the combination of the crucifixion and the rupture of the heart.

That rupture of the heart actually occurred I firmly believe."

5. WAS A DECEPTIVE ILLUSION TRICK USED?

Andre Kole is considered one of the world's leading illusionists, often called the magician's magician. He has never been fooled by another illusionist or magician. He has created and sold more than 1,400 illusionary and magical effects.

When Andre was a student, he studied psychology. He was challenged to apply his proficiency to the Resurrection, to explain it away by modem magic and illusion. He accepted the challenge – but concluded that there is no way through modern illusionary effects or magic that Jesus could have deceived His followers.

Once, when discussing this with me, he said, "Josh, there are too many built-in safety factors." Consider the weight of the two-ton stone rolled against the tomb, the fear of death for the Roman guards if they failed in their duty, the physical state of a crucified man, to name a few.

Kole was forced to the conclusion that if the Resurrection was a lie, the disciples must have known it was a lie.

6. DECEIT BY THE DISCIPLES?

If the disciples lied about the Resurrection, then they died for a lie. Good historical tradition shows us twelve Jewish men, eleven of whom died martyrs' deaths as a tribute to one thing: an empty tomb and the appearances of Jesus of Nazareth alive after His death by crucifixion.

Remember that at first the disciples didn't believe it either – not until they saw Him with their own eyes. For forty days after His resurrection, these men walked with Jesus, lived with Him, ate with Him. His resurrection was accompanied by many "convincing proofs" (Acts 1:3).

While it's true that thousands of people throughout history have died for a lie, they did so *only* if they thought it to be the truth. Tertullian said, "No man would be willing to die unless he knew he had the truth."[8]

What happened to these disciples of Jesus? Dr. Michael Green points out that:

"The Resurrection was the belief that turned brokenhearted followers of a rabbi into the courageous witness and martyrs of the early church... You could imprison them, flog them, but you could not make them deny their conviction that 'on the third day, he rose.'"

CHRISTIANITY HOAX OR HISTORY?

CONSIDER THE FACTS

So many security precautions were taken with the trial, crucifixion, burial, entombment, sealing, and guarding of Christ's tomb that it becomes very difficult for critics to defend their position that Christ did not rise from the dead.

FACT #1: BROKEN ROMAN SEAL

The first obvious fact is the breaking of the seal that stood for the power and authority of the Roman Empire. The consequences of breaking the seal were extremely severe.

Once the seal was violated, the "FBI" of the Roman Empire was called into action to find the person or persons who were responsible. If they were apprehended, it meant automatic execution by crucifixion upside down (where your guts ran into your throat). People feared the breaking of the seal.

The disciples after the crucifixion of Jesus were an unlikely group to risk such an act. They were afraid for their lives. Remember that even before the Crucifixion, when Jesus was arrested in the Garden of Gethsemane, they left Him and ran away. Peter denied that he knew

Jesus three times in one night. Only John and some of the women were with Jesus when He died. They spent the next few days behind closed doors "for fear of the Jews" (John 20:19).

FACT #2: EMPTY TOMB

Another obvious fact was the empty tomb. The disciples of Jesus did not flee to Athens or Rome to preach that Christ was raised from the dead. Rather, they went right back to Jerusalem, where, if their claims were false, the falsity would be evident.

The empty tomb was "too notorious to be denied." The burial site was well known not only to Christians and Jews but also to the Romans. This is why Dr. Paul Althaus states that the Resurrection "could not have been maintained in Jerusalem for a single day, for a single hour, if the emptiness of the tomb had not been established as a fact for all concerned."[1]

Both Jewish and Roman sources and traditions admit an empty tomb. Those sources range from Josephus to a compilation of fifth-century Jewish writings called the *Toledoth Jeshu*. Even the Jewish leaders acknowledged that the tomb was empty. Dr. Paul Maier calls this "positive evidence from a hostile source, which is the strongest kind of historical evidence. In essence, this means that if a source admits a fact decidedly not in its favor, then that fact is genuine."

Please keep in mind that the earliest Jewish reaction to the proclamation of Christ's resurrection was an aggressive attempt to *explain away* the empty tomb, not deny that it was empty. Dr. Ron Sider puts it this way: "If the Christians and their Jewish opponents both agree that the tomb was empty, we have little choice but to accept the empty tomb as historical fact."

Dr. Maier observes that "if all the evidence is weighed carefully and fairly, it is indeed justifiable, according to the canons of historical research, to conclude that the sepulcher of Joseph of Arimathea, in which Jesus was buried, was actually empty on the morning of the first Easter. And no shred of evidence has yet been discovered in literary sources, epigraphy, or archaeology that would disprove this statement." Dr. D. H. Van Daalen concludes that "it is extremely difficult to object to the empty tomb on historical grounds."

FACT #3: Large Stone Moved

On that Sunday morning the first thing that impressed the people who approached the tomb was the unusual position of the one-and-a-half- to two-ton stone that had been lodged in front of the doorway. All the Gospel writers mention it: The stone had been rolled away – not just away from the entrance to the tomb, but away from the tomb itself.

Now, I ask you, if the disciples had wanted to tiptoe around the sleeping guards, roll the stone away, and steal Jesus' body, how could they have done that without the guards' awareness? Those soldiers, even if asleep, would have to have had cotton in their ears, with earmuffs on, along with a heavy dose of knockout pills, not to have heard that huge stone being moved.

FACT #4: Roman Guard Goes Awal

The Roman guards fled. They left their place of responsibility. How can their dereliction of duty be explained, when Roman military discipline was so exceptional?

The Justinian Code, compiled in the sixth century, mentions in *Digest No. 49* all the offenses that required the death penalty under Roman law. The fear of their superiors' wrath and the possibility of death meant that Roman soldiers paid close attention to the most minute details of their job. Falling asleep on duty, leaving one's position, and failing in any way resulted in severe discipline.

One way a guard was put to death was by being stripped of his clothes and then burned alive with a fire started with his garments. If it was not apparent which soldier had failed in his duty, then lots were drawn to see which one would be punished with death for the guard unit's failure.

Certainly the entire unit would not have fallen asleep with that kind of threat over their heads. Dr. George Currie, a student of Roman military discipline, wrote that fear of punishment "produced flawless attention to duty, especially in the night watches."[2]

Dr. Bill White is in charge of the Garden Tomb in Jerusalem. His responsibilities have caused him to study the Resurrection and subsequent events. Dr. White makes several observations about the fact that the Jewish authorities bribed the Roman guards to say that Jesus' disciples had stolen His body:

> If the stone were simply rolled to one side of the tomb, as would be necessary to enter it, then they might be justified in accusing the men of sleeping at their posts, and in punishing them severely.

> If the men protested that the earthquake broke the seal and that the stone rolled back under the vibration, they would still be liable to punishment for behavior that might he labeled cowardice. But these possibilities do not meet the case.

> There was some undeniable evidence which made it impossible for the chief priests to bring any charges against the guard.

> The Jewish authorities must have visited the scene, examined the stone, and recognized its position as making it humanly impossible for their men to have permitted its removal.

> No twist of ingenuity could provide an adequate answer or a scapegoat, and so they were forced to bribe the guard and seek to hush things up.

FACT #5: GRAVE CLOTHES TELL A TALE

In a literal sense, against all reports to the contrary, the tomb was not totally empty – because of an amazing phenomenon.

After visiting the grave and seeing the stone rolled away, the women ran back and told the disciples. Then Peter and John took off running. John outran Peter and upon arriving at the tomb did not enter. Instead, he leaned over, looked in, and saw something so startling that he immediately believed.

He looked over to the place where the body of Jesus had lain, and there were the grave clothes, in the form of the body, slightly caved in and empty-like the empty chrysalis of a caterpillar's cocoon. That's enough to make a believer out of anybody. John never did get over it.

The first thing that stuck in the minds of the disciples was not the empty tomb but the empty grave clothes.

FACT #6: JESUS'S APPEARANCES CONFIRMED

Christ appeared on several occasions after the cataclysmic events of that first Easter.When studying an event in history, it is important to know whether enough people who were participants or eyewitnesses to, the event were alive when the facts about the event were published. To know this is obviously helpful in ascertaining the accuracy of the published report.

If the number of eyewitnesses is substantial, the event can be regarded as fairly well established. For instance, if we all witness a murder, and a later police report turns out to be a fabrication of lies, we, as eyewitnesses, can refute it.

a. There Were More Than 500 Witnesses

Several important factors are often overlooked when considering Christ's post-resurrection appearances to individuals. The first is the large number of witnesses who saw Him after that resurrection morning.

One of the earliest records of Christ's appearing after the Resurrection is by Paul in his letter to the Corinthians. The apostle appealed to their knowledge of the fact that Christ had been seen by more than 500 people at one time. Remember, as Paul emphasized, the majority of those people were still alive and could be questioned.

Dr. Edwin M. Yamauchi, associate professor of history at Miami University in Oxford, Ohio, emphasizes:

> What gives a special authority to the list [of witnesses] as historical evidence is the reference to most of the five hundred brethren being still alive. St. Paul says in effect, "If you do not believe me, you can ask them." Such a statement in an admittedly genuine letter written within thirty years of the event is almost as strong evidence as one could hope to get for something that happened nearly two thousand years ago."

> Let's take the more than 500 witnesses who saw Jesus alive after His death and burial and place them in a courtroom. Do you realize that if each of those 500 people were to testify for only six minutes, including cross examination, you would have an amazing fifty hours of firsthand testimony? Add to this the testimony of many other eyewitnesses and you could well have the largest and most lopsided trial in history.

b. There Were a Variety of Witnesses

Another factor often overlooked is the variety of situations and people to whom Jesus appeared. Merrill C. Tenney, former professor at Wheaton College, writes:

> It is noteworthy that these appearances are not stereotyped. No two of them are exactly alike. The appearance to Mary Magdalene occurred in early morning (John 20:1); to the travelers to Emmaus in the afternoon (Luke 24:29); and to the apostles in the evening, probably after dark (Luke 24:36). He

appeared to Mary in the open air (John 20:14) [but to the disciples in a closed room (John 20:19)]. Mary was alone when she saw Him; the disciples were together in a group; and Paul records that on one occasion he appeared to more than five hundred at a time (I Corinthians 15:6).

The reactions were also varied. Mary was overwhelmed with emotion (John 20:16-17); the disciples were frightened (Luke 24:37); Thomas was obstinately incredulous when told of the Lord's resurrection (John 20:25), but worshiped Him when He manifested Himself. Each occasion has its own peculiar atmosphere and characteristics, and revealed some different quality of the risen Lord.

c. There Were Hostile Witnesses

A third factor crucial to interpreting Christ's appearances is that He also appeared to those who were hostile or unconvinced. Over and over again I have read or heard people comment that Jesus was seen alive after His death and burial only by His friends and followers. But that line of reasoning is so pathetic it hardly deserves comment.

No author or informed individual would regard Saul of Tarsus as being a follower of Christ. The facts show the exact opposite. Saul despised Christ and persecuted His followers. It was a life-shattering experience when Christ appeared to him on the Damascus road. Although he was not at that time a disciple, he later became the Apostle Paul, one of the greatest witnesses for the truth of the Resurrection.

Also consider James, the brother of Jesus (not James the apostle and elder brother of John). History indicates that Jesus' brother was anything but a believer (John 7:3-5). Yet James not only became a follower of his brother but also died a martyr's death. What caused that change in his attitude and eventually his life?

According to his presence with the followers of Jesus as mentioned in Acts 1:13, his conversion must have occurred very shortly after Jesus' resurrection. The only historical explanation is what Paul said in 1 Corinthians 15:7--Jesus had appeared to James.

The argument that Christ's appearances were only to followers is an argument for the most part from silence, and arguments from silence can be dangerous. It is correct to say that all to whom Jesus appeared eventually became a follower. This is perhaps the best explanation of the conversion of so many of the Jerusalem priests (Acts 6:7).

JESUS CHRIST AND HIS RESURRECTION ARE HISTORICAL FACTS

Professor Thomas Arnold, for fourteen years a headmaster of Rugby, author of the famous *History of Rome*, and appointed to the chair of modern history at Oxford, was well acquainted with the value of evidence in determining historical facts.

This great scholar said:

> I have been used for many years to study the histories of other times, and to examine and weigh the evidence of those who have written about them, and I know of no one fact in the history of mankind which is proved by better and fuller evidence of a fair inquirer, than the great sign which God hath given us that Christ died and rose again from the dead.[3]

Brooke Foss Westcott, an English scholar, said:

> Taking all the evidence together, it is not too much to say that there is no historic incident better or more variously supported than the resurrection of Christ. Nothing but the antecedent assumption that it must be false could have suggested the idea of deficiency in the proof of it.

One man who was highly skilled at dealing with evidence was Dr. Simon Greenleaf. He was the famous Royal Professor of Law at Harvard University and succeeded Justice Joseph Story as the Dane Professor of Law in the same university.

Greenleaf examined the value of the historical evidence for the resurrection of Jesus Christ to ascertain the truth. He applied the principles contained in his three-volume treatise on evidences. He came to the conclusion that, according to the laws of legal evidence used in courts of law, there is more evidence for the historical fact of the resurrection of Jesus Christ than for just about any other event in ancient history.

REAL PROOF - THE DISCIPLES' LIVES

But the most telling testimony of all must be the lives of those early Christians. We must ask ourselves: What caused them to go everywhere telling the message of the risen Christ?

Had there been any visible benefits accruing to them from their efforts – prestige, wealth, increased social status, or material benefits – we might logically attempt to account for their actions, for their wholehearted and total allegiance to this "risen Christ. "

As a reward for their efforts, however, those early Christians were beaten, stoned to death, thrown to the lions, tortured, crucified. Every conceivable method was used to stop them from talking.

Yet they were peaceful people. They forced their beliefs on no one. Rather, they laid down their lives as the ultimate proof of their complete confidence in the truth of their message.

It has been rightly said that they went through the test of death to determine their veracity. It is important to remember that initially the disciples didn't believe. But once convinced – in spite of their doubts – they were never to doubt again that Christ was raised from the dead.

Do you know the odds of twelve men, all knowing something was a lie, not cracking under the torture and pressure to admit their deception?

AN EXAMPLE OF A CONSPIRACY CAVE-IN

Charles Colson, of Watergate scandal fame, writes that the Watergate cover-up revealed the true nature of humanity under pressure--the survival instinct. Ironically, his learning as an attorney and his years of experience in politics convinced him that Watergate demonstrates that the resurrection of Christ must be true.

This is how Colson arrived at his conclusion: A "thinly disguised panic began to sweep the plush offices of the stately old building that houses the most influential and powerful men in the world."

Yet he saw *that even "with the most powerful office in the world at stake, a small band of hand-picked loyalists, no more than ten of us, could not hold a conspiracy together for more than two weeks.* Think of the power at our fingertips: A mere command from one of us could mobilize generals and cabinet officers, even armies; we could hire or fire personnel and manage billions in agency budgets."

But yet with all this power, prestige, and their personal reputations and the luxury of their offices at stake, this group of men could not contain a lie.

However, Colson asks: "Was the pressure really all that great at that point? There had certainly been moral failures, criminal violation, even perjury by some. There was certain to be keen embarrassment; at the worst, some might go to prison, though that possibility was by no means certain. But no one was in grave danger; no one's life was at stake.

"Yet, after just a few weeks," observes Colson, "the natural human instinct for self-preservation was so overwhelming that the conspirators, one by one, deserted their leader, walked away from their cause, turned their backs on the power, prestige, and privileges."

How does all this relate to the Resurrection? One criticism of the veracity of Christ's resurrection is that His twelve disciples conceived a "Passover plot." They secretly stole away the body of Christ and neatly disposed of it, and then to their dying breaths maintained a conspiratorial silence. Colson concludes that...

> If one is to assail the historicity of the Resurrection and therefore the deity of Christ, *one must conclude that there was a conspiracy--a cover-up if you will--by eleven men with the complicity of up to five hundred others.*
>
> To subscribe to this argument, one must also be ready to believe that each disciple was willing to be ostracized by friends and family, live in daily fear of death, endure prisons, live penniless and hungry, sacrifice family, be tortured without mercy, and ultimately die--all without ever once renouncing that Jesus had risen from the dead!

This is why the Watergate experience is so instructive to me. If John Dean and the rest of us were so panic-stricken, not by the prospect of beatings and execution, but by political disgrace and possible prison term, one can only speculate about the emotions of the disciples.

Unlike the men in the White House, the disciples were powerless people, abandoned by their leader, homeless in a conquered land. Yet they clung tenaciously to their enormously offensive story that their leader had risen from His ignoble death and was alive – and was the Lord.

The Watergate cover-up reveals, I think, the true nature of humanity. None of the memoirs suggest that anyone went to the prosecutor's office out of such noble notions as putting the Constitution above the president or bringing rascals to justice--or even moral indignation.

Instead, *the writings of those involved are consistent recitations of the frailty of men. Even political zealots at the pinnacle of power will save their own necks in the crunch, though it may be at the expense of the one they profess to serve so zealously.*

Is it really likely, then, that a deliberate cover-up, a plot to perpetuate a lie about the Resurrection, could have survived the violent persecution of the apostles, the scrutiny of early church councils, the horrendous purge of the first-century believers who were cast by the thousands to the lions for refusing to renounce the lordship of Christ?

Is it not probable that at least one of the apostles would have renounced Christ before being beheaded or stoned? Is it not likely that some "smoking gun" document might have been produced exposing the "Passover plot?" Surely one of the conspirators would have made a deal with the authorities. Government and Sanhedrin probably would have welcomed such a soul with open arms and pocketbooks!

Take it from one who was inside the Watergate web looking out, who saw firsthand how vulnerable a cover-up is: Nothing less than a witness as awesome as the resurrected Christ could have caused those men to maintain to their dying day that Jesus is alive and Lord.

The weight of evidence tells me the apostles were indeed telling the truth.

CHRISTIANITY HOAX OR HISTORY?
THE RECORD PRESERVED

After a "free-speech" outdoors lecture I gave at Arizona State University, a professor accompanied by students from his graduate seminar on world literature approached me and said, "Mr. McDowell, you are basing all your claims about Christ on a second-century document that is obsolete. I showed in class today how the New Testament was written so long after Christ that it could not be accurate in what it recorded."

"Sir," I replied, "your opinions about the New Testament are twenty-five years out of date."

I knew where this professor and his students were coming from. As a university student, I had set out to prove that the New Testament was a collection of myths, half-truths, and outright errors. Instead, I ended up with historical evidence for the Bible's reliability that was overwhelming. If other literature of antiquity had the same historical evidence, no one would question its authenticity and reliability.

"So, who cares?" you say. You do. To one degree or another you have developed an opinion on the reliability of the New Testament and its application to your own life. Maybe you haven't thought much about it and just ignore the implications. Maybe you feel skeptical because it was written a long time ago-what possible relevance could it have today? Maybe all those "miracles"-and to top it off, the Resurrection – disqualify it in your mind for serious study. Or maybe you want to believe, but it seems so full of contradictions.

Are you willing to talk about it and look at the facts? Good. Me, too.

QUESTION 1

How can the New Testament accurately report the facts about Jesus if it wasn't written until 100 years later?

Many opinions about the records concerning Jesus are based on the conclusions of F. C. Baur, a German critic. Baur assumed that most of the New Testament Scriptures were not written until late in the second century A.D. He concluded that these writings came basically from myths or legends that had developed during the lengthy interval between the lifetime of Jesus and the time those accounts were set down in writing.

FACT: *Recent archaeological discoveries point to the first-century origin of New Testament manuscripts.*

FACT: *There is strong evidence within the New Testament that it was written at an early date.*

The Book of Acts records the missionary activity of the early church and was written as a sequel by the same person who wrote the Gospel according to Luke. The Book of Acts ends with the Apostle Paul being alive in Rome. This leads us to believe that it was written before he died, since the other major events of his life were recorded. There is reason to believe that Paul was put to death in Nero's persecution of Christians in A.D. 64, which means the Book of Acts was composed before then.

The death of Christ took place around A.D. 30. If the Book of Acts was written before A.D. 64, then the Gospel of Luke was written sometime in the intervening thirty years.

The early church generally taught that the first Gospel composed was Matthew, which places it still closer to the time of Christ. This evidence leads us to believe that the first three Gospels were composed within thirty years of the time these events occurred, when unfriendly witnesses were still living who could have contradicted the Gospels if they had not been accurate.

QUESTION 2

But aren't the New Testament stories just a bunch of myths and legends that finally got written down?

Some critics argue that information about Christ was passed by word of mouth until it was written down in the form of the Gospels. Even though the period was much shorter than previously believed, they conclude that the Gospel accounts took on the forms of tales and myths.

FACT: *The period of oral tradition is not long enough to allow for the development of myths and legends.*

Dr. Simon Kistemaker, who has studied the development of myths and legends wrote:

> "Normally the accumulation of folklore among people of primitive culture takes many generations; it is a gradual process spread over centuries of time. But . . . we must conclude that the Gospel stories were produced and collected within little more than one generation."

Professor A. N. Sherwin-White, a prominent historian of Roman/ Greek times, points out that for the New Testament accounts to be legend, the rate of legendary accumulation would have to be unbelievably accelerated; more generations are needed.

QUESTION 3

How do we know that the Bible we read today is the same as when it was originally written?

In other words, since we don't have the original documents, how do we know the copies we have are reliable? Accusations abound about zealous monks changing the biblical text as it was copied during the Dark Ages.

FACT: *Although we do not possess originals, copies exist from a very early date.*

When I first wrote *Evidence That Demands a Verdict*, I was able to document 14,000 manuscripts of the New Testament. However, with new discoveries, I can document 24,633 manuscripts of just the New Testament. Altogether there are more than 24,000 New Testament manuscripts and portions thereof in Greek and other early versions!

The significance of this number of manuscripts documenting the New Testament is even greater when one realizes that in all of ancient history, the second runner-up in terms of manuscript authority is the *Iliad* by Homer--and it has only 643 surviving documents.

FACT: *The time span between the originals and the earliest copies in possession is extremely short.*

The New Testament was originally written in Greek. Though we do not have any originals, there are approximately 5,500 Greek copies in existence that contain all or part of the New Testament in Greek. The earliest fragment dates about A.D. 120. Two major manuscripts, Codex Vaticanus (A.D. 325) and Codex Sinaiticus (A.D. 350), a complete copy of the New Testament, date within 250 years of the original writing. That may seem like a long time span, but it is minimal compared to most ancient works. The first complete copy of the *Odyssey* is from 2,200 years after it was written!

Documentary Evidences of Ancient Manuscripts

Author/Work	When Written	Earliest Copy	Time Span	Number Of Copies
Caesar	100-44 BC	AD 900	1,000 yrs.	10
Livy	59 BC	AD 1700	1,750 yrs	20
Plato (Tetralogies)	427-347 BC	AD 900	1,200 yrs.	7
Tacitus (Annals)	AD 100	AD 1100	1,000 yrs.	20
Tacitus (minor works)	AD 100	AD 1000	900 yrs.	1
Pliny the Younger (History)	AD 61-113	AD 850	750 yrs.	7
Thucydides (History)	460-400 BC	AD 900	1,300 yrs.	8
Suetonius (De Vita Caesarum)	AD 75-160	AD 950	800 yrs.	8
Herodotus (History)	480-425 BC	AD 900	1,300 yrs.	8
Sophocles	496-406 BC	AD 1000	1,400 yrs.	193
Lucretius	55 or 53 BC	AD 1050	1,100 yrs.	2
Catullus	54 BC	AD 1550	1,600 yrs.	3
Euripides	480-406 BC	AD 1100	1,500 yrs.	9
Demosthenes	383-322 BC	AD 1100	1,300 yrs.	200*
Aristotle	384-322 BC	AD 1100	1,400 yrs.	49**
Aristophanes	450-385 BC	AD 900	1,200 yrs.	10
Homer (Iliad)	900 BC	400 BC	500 yrs.	643
New Testament	AD 40-100	AD 125	**25 yrs.**	**24,000-plus**

From Evidence That Demands a Verdict, 42-43.

* All from one copy.

** Of any one work.

A few years ago, 36,000 *quotations* of the Scriptures by the early church fathers could be documented. But more recently, as a result of research done at the British Museum, we are now able to document 89,000 quotations from the New Testament in early church writings. If you destroyed all the Bibles and biblical manuscripts, one could reconstruct all but eleven verses of the entire New Testament from quotations found in other materials written within 150 to 200 years after the time of Jesus Christ!

These facts are called the *bibliographical* test, which determines only that the text we have now is what was originally written.

QUESTION 4

How do we know the writers got their facts straight in the first place? Maybe it was just hearsay.

"Hearsay" is not admissible as evidence in a court of law. *The Federal Rules of Evidence* declares that a witness must testify concerning what he has firsthand knowledge of, not what has come to him indirectly from other sources.

FACT: *The New Testament does not fit the mode of hearsay.*

Concerning the value of a person testifying of his own knowledge, Dr. John Warwick Montgomery, an attorney and dean of the Simon Greenleaf School of Law, points out that from a legal perspective, the New Testament documents meet the demand for "primary-source" evidence. He writes that the New Testament record is"

> "Fully vindicated by the constant assertions of their authors to be setting forth that which we have heard, which we have seen with our eyes, which we have looked upon and our hands have handled."[2]

FACT: *Most testimony in the New Testament comes from firsthand knowledge.*

For example, when Mary went to the tomb, the angel appeared to her and said, "He is not here, He has risen." When Mary told the disciples, it was hearsay because she hadn't seen Him herself; she just had heard about it. But later, Jesus personally appeared to Mary. That took it out of hearsay and made her testimony a primary source.

Dr. Louis Gottschalk, former professor of history at the University of Chicago, outlines his historical method in an excellent guide used by many for historical investigation. Gottschalk points out that the ability of the writer or the witness to tell the truth is helpful to the historian to determine credibility, "even if it is contained in a document obtained by force or fraud, or is otherwise impeachable, or is based on hearsay evidence, or is from an interested witness."[3]

This ability to tell the truth, Gottschalk points out, is closely related to the witness's nearness both geographically and chronologically to the events recorded.

What about the New Testament accounts? The New Testament accounts of the life and teachings of Jesus were recorded by men who either had been eyewitnesses themselves or who were recounting the descriptions of eyewitnesses. For instance:

- Luke wrote to Theophilus, "It seemed fitting for me as well, having investigated everything carefully from the beginning, to write it out for you in consecutive order" (Luke 1:1-3).
- Peter wrote, "We were eyewitnesses" (2 Peter 1:16).
- Wrote John, "What we have seen and heard we proclaim to you . . ." (1 John 1:3) and "his witness is true, and he knows that he is telling the truth ..." (John 19:35).
- Luke painstakingly listed proven historical facts (Luke 3:1).

This closeness to the recorded accounts is an extremely effective means of certifying the accuracy of what is retained by a witness.

QUESTION 5

But what if the writers simply told falsehoods?

Good question. The historian does have to deal with the eyewitness who consciously or unconsciously tells falsehoods, even though he is near the event and is competent to tell the truth.

FACT: *The New Testament writers appealed to common knowledge about Jesus.*

The New Testament accounts of Christ were being circulated within the lifetimes of His contemporaries. Those people could have confirmed or denied the accuracy of the accounts. The writers not only said, "Look, we saw this" or "We heard that."

But right in front of their most severe opponents they turned the tables around and said, "You also know about these things--you saw them yourselves." (One had better be careful when he says to the opposition, "You know this also," because if he isn't right in the details, he will be exposed immediately!)

Speaking to the Jewish people, Peter said, "Men of Israel, listen to these words: Jesus the Nazarene, a man attested to you by God with miracles and wonders and signs which God performed through Him" [notice this] *"in your midst, just as you yourselves know . . . ,"* (Acts 2:22). If they hadn't seen those miracles for themselves, Peter never would have gotten out of there alive, let alone have thousands trust in Christ.

F. F. Bruce, a professor at Manchester University, makes an astute observation in his book *The New Testament Documents--Are They Reliable?* about the value not only of friendly witnesses (those that agree with you), but also *hostile witnesses:* "The disciples could not afford to risk inaccuracies (not to speak of willful manipulation of the facts) which would at once be exposed by those who would only have been glad to do so."[4]

QUESTION 6

So Jesus died on the cross, and later His followers were killed. But "dying for a great cause" doesn't prove the truth of that cause, does it? After all, a lot of people in history have died for great causes.

FACT: *What the disciples thought was their "great cause" died on the cross.*

When Jesus died that Friday, the disciples no longer had a "great cause." Remember, the Jews at that time were under oppression from the Romans. To hold the allegiance of the people, the

Jewish leaders taught that when the Messiah came, He would come as a reigning political Messiah, and He'd throw the Romans out.

That is why it was so hard for the apostles to understand what Jesus was saying. He said, "I have to die. I have to go to Jerusalem. I'm going to be crucified and buried." They couldn't understand it. Why? From childhood it had been ingrained into them that when the Messiah came, He would reign politically. They thought they were in on something big. They were going to rule with Him.

Professor E. F. Scott, in his book *Kingdom and the Messiah*, points out that:

> "For the people at large, their Messiah remained what He had been to Isaiah and his contemporaries, the Son of David, who would bring victory and prosperity to the Jewish nation."[5]

Dr. Jacob Gardenhus, a Jewish scholar, observed that the Jews awaited the Messiah as the One who would deliver them from Roman oppression. The temple with its sacrificial service was intact, because the Romans did not interfere in Jewish religious affairs. The messianic hope was basically for national liberation, for a Redeemer of a country that was being oppressed.

The Jewish Encyclopedia records that the Jews "yearned for the promised Deliverer of the house of David who would free them from the yoke of the hated foreign usurper, who would put an end to the impious world and rule, and would establish His own reign of peace and justice in its place."[6]

Therefore, at the point of Jesus' crucifixion, the disciples "great cause" was dead from their natural perspective. There would have been nothing for them to die for. Their hopes were dashed.

FACT: *It was the Resurrection that totally changed the lives of the disciples.*

But then something happened. In a matter of a few days their lives were turned upside down. All but one became a martyr for the cause of the Man who appeared to them after His death. With the Resurrection they finally understood what Jesus had been saying: He had come to suffer and die for the sins of the world, and He would come a second time to reign throughout the world.

The Resurrection is the only thing that could have changed those frightened, discouraged disciples into apostles who would dedicate their lives to spreading His message. Once they were convinced of it, they never denied it.

QUESTION 7

Isn't the Bible just witnessing to itself?

Okay, so the "internal evidence" is pretty convincing that the New Testament picture of Christ can be trusted. But isn't that just the Bible being its own witness? Are there any other sources of proof?

FACT: *At least two historians of the time offer external evidence as well.*

The historian Eusebius preserves some writings of Papias, bishop of Hierapolis (A.D. 130):

> The Elder [Apostle John] used to say this also: "Mark, having been the interpreter of Peter, wrote down accurately all that he [Peter] mentioned, whether sayings or doings of Christ, not, however, in order. For he was neither a hearer nor a companion of the Lord; but afterward, as I said, he accompanied Peter, who adapted his teachings as necessity required, not as though he were making a compilation of the sayings of the Lord. So then Mark made no mistake, writing

down in this way some things as he mentioned them; for he paid attention to this one thing, *not to omit anything that he had heard, nor to include any false statement among them*" (emphasis added).

Another historian, Irenaeus, bishop of Lyons (A.D. 180), preserves the writings of Polycarp, bishop of Smyrna, who had been a Christian for eighty-six years and was a disciple of John the apostle:

> So firm is the ground upon which these Gospels rest, that the very heretics themselves bear witness to them, and, starting from these, each one of them endeavors to establish his own particular doctrine.

Polycarp was saying that the four Gospel accounts about what Christ said and did were so accurate (firm) that even the heretics themselves in the first century could not deny their record of events. Instead of attacking the scriptural account, which would have proven fruitless, the heretics started with the teachings of Jesus and developed their own heretical interpretations. Since they weren't able to say, "Jesus didn't say that," they instead had to say, "This is what He meant..." (You are on pretty solid ground when you get those who disagree with you to do that!)

FACT: *Archaeology, too, often provides powerful external evidence.*

Archaeology contributes to biblical criticism, not in the area of inspiration and revelation, but by providing evidence of accuracy about events that are recorded. Archaeologist Joseph Free, in his book, *Archaeology and Bible History*, says that archaeology has confirmed countless biblical passages that were earlier rejected by critics as unhistorical or contradictory to supposedly "known" facts."[7]

For instance, Luke at one time was considered incorrect for referring to the Philippian rulers as *praetors*. According to the "scholars," two *duumuirs* would have ruled the town. However, Luke was right.

Archaeological findings have shown the title of *praetor* was employed by the magistrates of a Roman colony.

Luke's choice of the word *proconsul* as the title for Gallio also has been proven correct, as evidenced by the Delphi inscription which states: "As Lucius Junius Gallio, my friend, and the proconsul of Achaia ..." (compare Acts 18:12).

Again and again Luke's historical references have been substantiated. Notice that in the first verse of Luke 3 there are fifteen historical references given by Luke that can be checked for accuracy:

> "Now in the fifteenth year [that's one historical reference] of the reign of Tiberius Caesar [that's two], when Pontius Pilate [three] was governor [four] of Judea [five], and Herod [six] was tetrarch [seven] of Galilee [eight], and his brother Philip [nine] was tetrarch [ten] of the region of Ituraea and Trachonitis [eleven and twelve], and Lysanias [thirteen] was tetrarch [fourteen] of Abilene [fifteen] ..."

It is no wonder that E. M. Blaiklock, professor of classics at Auckland University, concludes that:

> "Luke is a consummate historian, to be ranked in his own right with the great writers of the Greeks."[8]

FACT: *One test of a writer is consistency.*

Commenting on the overall historical accuracy of Luke, F. F. Bruce (noted earlier) says:

> "A man whose accuracy can be demonstrated in matters where we are able to test it is likely to be accurate even where the means for testing him are not available ... Luke's record entitles him to be regarded as a writer of habitual accuracy."[9]

FACT: *The same standard or test should be applied to the Bible as is applied to secular literature.*

There was a time in my life when I myself tried to shatter the historicity and validity of the Scriptures. But I have come to the conclusion that they are historically trustworthy. If a person discards the Bible as unreliable in this sense, then he or she must discard almost all the literature of antiquity.

One problem I constantly face is the desire on the part of many to apply one standard or test to secular literature and another to the Bible. But we need to apply the same test, whether the literature under investigation is secular or religious, without incorporating presuppositions or assumptions that rule out certain content, i.e., the supernatural.

Dr. Clark Pinnock, in his book *Set Forth Your Case*, concluded after extensive research:

> "There exists no document from the ancient world, witnessed by so excellent a set of textual and historical testimonies and offering so superb an array of historical data on which an intelligent decision may be made. An honest person cannot dismiss a source of this kind. Skepticism regarding the historical credentials of Christianity is based upon an irrational bias."[10]

F. F. Bruce makes the following observation:

> The evidence for our New Testament writings is ever so much greater than the evidence for many writings of classical authors, the authenticity of which no one dreams of questioning ...[11]

And if the New Testament were a collection of secular writings, their authenticity would generally be regarded as beyond all doubt.

FACT: *The New Testament portrays historical reality.*

The late historian Will Durant, trained in the discipline of historical investigation" who spent his life analyzing records of antiquity, writes:

> Despite the prejudices and theological preconceptions of the evangelists, they record many incidents that mere inventors would have concealed - the competition of the apostles for high places in the Kingdom, their flight after Jesus' arrest, Peter's denial, the failure of Christ to work miracles in Galilee, the references of some authors to His possible insanity, His despairing cry on the cross; no one reading these scenes can doubt the reality of the figure behind them.
>
> That a few simple men should in one generation have invented so powerful and appealing a personality, so lofty an ethic, and so inspiring a vision of human brotherhood, would be a miracle far more incredible than any recorded in the Gospels.
>
> After two centuries of Higher Criticism, the outlines of the life, character, and teachings of Christ remain reasonably clear, and constitute the most fascinating feature in the history of Western man.[12]

CHRISTIANITY HOAX OR HISTORY?

WHAT IT MEANS TODAY

What difference does it make if the Bible is historically accurate or not? After all, a lot of people regard the bible as good literature, like the works of Shakespeare or Aristotle.

But the historically accurate portrait of Christ in the New Testament has personal implications for everyone.

The claims that Scripture makes for itself (that it is the Word of God to us) and that Jesus makes for Himself (that He is God's Son, sent to redeem men and women and reconcile us to God) are either the biggest lies and the cruelest hoax foisted on the human race--or they are the most remarkable and noteworthy claims in history. The birth, life, death, and resurrection of Jesus was a turning point in the history of mankind. Measured by His influence, Jesus is central to the human story.

THE POWER OF CHRIST

The Christ of the New Testament can change lives. No matter what the critics say, the Christ of the New Testament changes lives. Millions from all backgrounds, nationalities, races, and professions, more than twenty centuries, are witnesses to the sin-breaking power of God's forgiveness through Jesus Christ.

E. Y. Mullins writes:

> I, too, am a walking testimony that the Scriptures are true, that Jesus Christ was raised from the dead and lives today. When I was a student, I set out to refute intellectually the Bible as a reliable document, the Resurrection as a factual historical event, and Christianity as a relevant alternative. After gathering the evidence, I was compelled to conclude that my arguments wouldn't stand up - that Jesus Christ is exactly who He claimed to be, the Son of God.

During my second year at the university, I became a Christian. You've probably heard religious people talk about their "bolt of lightning." Well, nothing so dramatic happened to me, but in time there were some very observable changes.

MENTAL PEACE

I had been a person who always had to be occupied. I had to be over at my girl's place or somewhere in a rap session. I'd walk across campus, and my mind would be a whirlwind of conflicts. I'd sit down and try to study or think, and I couldn't. But in the few months after I made the decision to trust Christ, a kind of mental peace began to develop. Don't misunderstand, I'm not talking about the absence of conflict. What I found in this relationship with Jesus wasn't so much the absence of conflict as it was the ability to cope with it. I wouldn't trade this for anything in the world.

CONTROL OF TEMPER

Another area that started to change was my bad temper. I used to "blow my stack" if somebody just looked at me cross-eyed. I still have the scars from almost killing a man my first year in the university. My temper was such an integral part of me, I didn't consciously seek to change it. Then one day after my decision to put my faith in Christ, I arrived at a crisis, only to find that my temper was gone!

Freedom from Resentment

I had a lot of hatred in my life. It wasn't something outwardly manifested, but there was a kind of inward grinding. I was ticked off with people, things, issues.

The one person I hated more than anyone else in the world was my father. I despised him. He was the town alcoholic. And if you're from a small town and one of your parents is an alcoholic, you know what I'm talking about. Everybody knew. My friends would come to high school and make jokes about my father. They didn't think it bothered me. I was laughing on the outside, but let me tell you, I was crying on the inside. I'd go out in the barn and find my mother lying in the manure behind the cows. She'd been knocked down by my father and couldn't get up.

About five months after I made my decision for Christ, love for my father – a love from God through Jesus Christ – inundated my life. It took that resentment and turned it upside down. It was so strong, I was able to look my father squarely in the eye and say, "Dad, I love you." I really meant it.

When I transferred to a private university, I was in a serious car accident. With my neck in traction, I was taken home. I'll never forget my father coming into my room, standing by my bed, and asking, "Son, how can you love a father like me?"

I said, "Dad, six months ago I despised you." Then I shared with him my conclusions about Jesus Christ and how He had changed me. Forty-five minutes later one of the greatest thrills of my life occurred. Somebody in my own family, someone who knew me so well I couldn't pull the wool over his eyes, my own father, said to me, "Son, if God can do in my life what I've seen Him do in yours, then I want to give Him the opportunity."

Usually changes take place over several days, weeks, or even years. But my father was changed right before my eyes. It was as though somebody reached in and turned on a light bulb. I've never seen such a rapid change before or since. My father touched alcohol only once after that He got it as far as his lips, and that was it. He didn't need it any more.

I've come to one conclusion: A relationship with Jesus Christ changes people. You can ignore Him; you can mock or ridicule Christianity. It's your decision. And yet, when all else is said and done, we must face the fact that Peter pointed out "Jesus [is] the Messiah... There is salvation in no one else! Under all heaven there is no other name for men to call upon to save them" (Acts 4:11-12).

If you ask Him to take control of your life, start watching your attitudes and actions – because the Christ of the new Testament is in the business of forgiving sin, removing guilt, changing lives, and building new relationships.

AUTHOR'S CLOSING COMMENTS
DON CARMICHAEL

In our present highly educated modern society, many consider the ideas of God, Holy Scriptures, Heaven and Hell to be old fashioned concepts that developed from folklore and myths. These religious notions, they feel, were created by simple-minded and uneducated people who had to construct a deity fantasy to make sense of their world and to give them hope.

But what about today? How can the sophisticated and enlightened person believe such things? An educated person focuses on science, not religion, right? After all, science is based on *facts*. It reveals *truth*.

Science does reveal many facets of truth. Science and Christianity are deeply compatible, and for the intellectually honest person, *science continues to provide increasing evidence and proofs of a divine Creator*. After all, He is the one who created all the laws of physics, chemistry, mathematics and biology upon which science is based.

Science merely observes these laws, but it can't create them. Your Heavenly Father, the Creator, is the one who established them and put them into motion.

And as sophisticated as science has become, it still can't sufficiently answer two very simple questions: *What is life* and *Where did it come from*? Science certainly can't create *life*, even with all of the world's most advanced technologies available to man.

Remember the story of when my wife and I encountered baby Reagan in the hospital, not realizing that he had already died? He had everything in his body package for life – perfect brain, heart, lungs, eyes, nervous system etc. – but where did his *life* go? What is *life*, and where does it come from? Again, science cannot adequately answer these simple questions.

The more we learn about the exquisite designs in the human body, animals and our natural world, the more overwhelming the evidence becomes of a brilliant *Intelligent Designer*. These are among the many "fingerprints" of God that are evident to those who seek truth.

Another set of God's "fingerprints" that can be clearly observed are the changed lives of the people who have a *genuine* relationship with Him. Did you catch that? Not a lip-service, *I-believe-in-God*-level of intellectual assent, which many profess but have little change in their outward lives, but a *genuine-relationship* level.

A genuine relationship with the Living God is *always transformational*. It *always* results in increased internal *peace, hope* and *joy* and in increased external *kindness* and *love* toward other people. Just in the small sample size of the encounter stories in this book, look at how many of these stories demonstrated dramatic before-and-after life changes after meeting Christ:

TED MELENDEZ - the violent biker and brutal gang enforcer who used to love to hurt people, had his bitter heart transformed and replaced with a *beyond*-natural love for people – to the point that people commented, "Your eyes have changed" and children call him "Uncle Teddy Bear." He also experienced an immediate and permanent release from drug and alcohol addictions.

JEREMIAH CASTILLE - when encountering his brother's killer face-to-face, was given the *beyond*-natural ability to forgive this murderer and have a genuine concern for his soul. And Jeremiah's mother, when she came to Christ in rehab, was immediately set free from alcohol addiction and stayed sober for the next 35 years to the end of her life.

JOSH MCDOWELL - who had such bitter hatred for his abusive alcoholic father that he would beat him horribly and tie him up in their farm's barn, who after coming to Christ, was able to forgive his father, hug him and say a heartfelt, "I love you!" This change in Josh's heart so touched his father, the town drunk, that he committed his life to Christ on the spot and experienced immediate release from decades of alcoholism. He, likewise, didn't take another drink for the rest of his life.

FRANK BARKER - the U.S. Navy fighter pilot who loved fast living, partying and sinning who, after coming to Christ, was filled with humility, a new love for people and a deep desire for pure living.

So, how about you? Are there observable "fingerprints" of God on your life that would show *clear evidence* – noticeable to other people – that you have a *genuine* relationship with Him?

Many people claim to "be a Christian" or to "know God" but don't. Many even do good works in God's name, including even many church leaders and pastors, who don't authentically know Him. No matter how good we are, our good works cannot save us.

Jesus, the Messiah of the Bible, speaks to this group of people and has some frightening words to say.

"Many will say to me on that day (the Day of Judgement), "Lord, Lord, did we not prophesy in your name, and in your name drive out demons, and in your name perform many miracles? Then I will tell them plainly, 'I never knew you. Depart from me, you cursed!'"
Matthew 7:22,23

A genuine relationship with God is built on *relationship*, not good works, lip service or intellectual acknowledgement that He exists. In the Bible, James comments about these people who believe in God, like so many people do today, but don't have a true and *saving relationship* with Him:

> *"You believe that there is one God? Good!*
> *Even the demons believe (that God exists) and shudder!"*
> *James 2:19*

Just "believing in God" doesn't count when it comes to salvation and eternal life.

Here are some questions to ask yourself to help determine whether or not you have an authentic, saving relationship with God – one that will ultimately lead to eternal life and Heaven:

• Do you have deep levels of *peace, hope* and *joy* in your heart, regardless of circumstances?

• Would the character qualities of *kindness, gentleness* and *patience* accurately describe you?

• Do you have an *increasing love* toward other people and *deepening concern* about their well-being?

• Do you desire *purity* in what you watch, say and do?

• Do you have a heart of love toward God and enjoy spending time with Him?

• Do you spend time in prayer and in His Word?

If your answers to these questions are no, you likely DO NOT have an authentic and saving relationship with God – even if you are a "good person," or describe yourself as a "Christian," or are even a regular church-goer. Your final eternal destination is likely not Heaven, and this should be a HUGE concern!

None of us knows when that final day will come in our lives when we die. No matter our age, none of us are promised tomorrow, and the end can come suddenly and unexpectedly.

Some of you reading this book are already in the twilight years of your lives, literally within a few months or few years of the end. Death is waiting for you around the corner. Are you ready to meet your Creator? You better be, because your *eternal soul* is at stake.

The real desire behind this book is to bring you into a life-changing relationship with God, your Heavenly Father, that will lead you to a more fulfilling life here and eternal life when you die. Life is short. Eternity is forever. The old-fashioned notions of Heaven and Hell are real, and one of those will be your eternal destination.

You are the one who chooses your destination

In fact, Jesus talked *far more* about Hell than Heaven! Imagine that! He implores you to choose life in Him, and He has patiently, throughout all of your life up to this day and this very moment, given you many opportunities to make this decision.

Your Heavenly Father, who formed you and loves you, is eager to have you accept His gift of salvation. His offer of eternal life is available to every person. He desires that *all* would enter into a transformational relationship with Him, and that *none* should perish:

"The Lord is not slow in keeping his promise, as some understand slowness.
Instead he is patient with you, not wanting anyone to perish,
but everyone to come to repentance."
2 Peter 3:9

"Behold, I stand at the door (to your heart) and knock.
If anyone hears my voice and opens the door, I will come in to him
and eat with him, and him with me."
Revelation 3:20

Conclusion

In conclusion, there are two epic scenes in the last book of the Bible that describe the two final eternal destinations awaiting all of humanity. Each man and woman chooses where he or she will ultimately spend their eternity.

It's their own personal *choice*.

One of these two locations is waiting for you.

Heaven

The destination for all those who have placed their trust in Jesus Christ (the Messiah), accepted His gift of eternal life and committed to follow Him as their Lord:

> *"And I heard a loud voice from the throne saying, "Look! God's dwelling place is now among the people, and he will dwell with them. They will be his people, and God himself will be with them and be their God. He will wipe every tear from their eyes. There will be no more death or mourning or crying or pain, for the old order of things has passed away."*
> *Revelation 21:3,4*

Hell

The destination for all those who have <u>chosen</u> NOT to place their trust in Jesus Christ (the Messiah), NOT to accept His gift of eternal life and NOT to follow Him as their Lord:

> *"Then I saw a great white throne and Him who was seated on it. The earth and the heavens fled from his presence, and there was no place for them. And I saw the dead, great and small, standing before the throne, and books were opened. Another book was opened, which is the Book of Life.*
>
> *The dead were judged according to what they had done as recorded in the*

books. The sea gave up the dead that were in it, and death and Hades gave
up the dead that were in them, and each person was judged according to
what they had done. Then death and Hades were thrown into
the lake of fire. The lake of fire is the second death.
Anyone whose name was not found written in the Book of Life
was thrown into the lake of fire."
Revelation 20:11-15

I hope you've enjoyed this book.

Get to know the One who formed you. He wants you to know Him and to satisfy the deepest longings of your heart with peace, hope, joy and abundant life – both here and in eternity.

For a reminder on how to enter into a saving relationship with God and gain eternal life, refer to Section 3 of this book, "God is Inviting You to a Personal Encounter."

For questions, advice or help, or to share any stories of
God's supernatural encounters in your life, email:
EncountersBookConversation@gmail.com

ABOUT THE AUTHOR

Don Carmichael is a native of Birmingham, Alabama and CEO of The Champion Group, a fundraising and organizational development consulting company serving schools and nonprofits in the U.S., Canada and overseas.

Over the years and often at critical and challenging moments in his life, he has personally experienced God's supernatural hand... *miracles*. He discovered that many others have had similar supernatural experiences, and thus this book was inspired.

Don is an avid reader, mentor and general life enthusiast. He serves on a variety of boards and is focused on a life of purpose centered on helping others. He and his wife, Vicki, reside on a small mini-farm and have four wonderful young-adult daughters.

For insights, advice or conversation about any aspects or topics
contained in this book, or to share any stories of
God's supernatural encounters in your life,
contact the author at:

EncountersBookConversation@gmail.com

APPENDIX
REFERENCES FROM SECTION FOUR - HOAX OR HISTORY?

INTRODUCTION

1. William E. Lecky, *History of European Morals from Augustus to Charlemagne* (New York: D. Appleton and Co., 1903), 2:8-9.

2. J. T. Fisher, and L. S. Hawley, *A Few Buttons Missing* (New York: MacMillan, 1947), 113.

Chapter 1: Back From The Grave

1. William F. Albright, *Recent Discoveries in Bible Lands* (New York: Funk and Wagnalls, 1955), 136.

2. Sir Frederick Kenyon, *The Bible and Archaeology* (New York: Harper and Row, 1940), 288-289.

3. Sir William Ramsay, *The Bearing of Recent Discovery on the Trustworthiness of the New Testament* (London: Hodder and Stoughton, 1915), 222.

4. F. F. Bruce, *The New Testament Documents: Are They Reliable?* rev. ed. (Grand Rapids, Mich.: Eerdmans, 1977), 33.

5. J. N. D. Anderson, Christianity: *The Witness of History* (Wheaton, Ill.: Tyndale, 1970), 92.

6.]ohn Warwick Montgomery, *History and Christianity* (Downers Grove, Ill.: InterVarsity Press, 1972), 78.

7. David Friedrich Strauss, *The Life of Jesus for the People*, 2d ed. (London: Williams and Norgate, 1879), 1:412.

8. Gaston F. Poote, *The Transformation of the Twelve* (Nashville, Tenn.: Abingdon, 1958), 12.

9. Michael Green, "Editor's Preface" in George Eldon Ladd, *I Believe in the Resurrection of Jesus* (Grand Rapids, Mich.: Eerdmans, 1975).

Chapter 2 - Consider The Facts

1. Paul Althaus, *Die Wahrheit des kirchlichen Osterglaubens* (Gutersloh: C. Bertelsmann, 1941), 22, 25ff.

2. George Currie, *The Military Discipline of the Romans from the Founding of the City to the Close of the Republic.* An abstract of a thesis published under the auspices of the Graduate Council of Indiana University, 1928, 41-43.

3. Thomas Arnold, *Christian Life--Its Hopes, Its Fears, and Its Close*, 6th ed. (London: T. Fellowes, 1859), 324.

Chapter 3 - The Record Preserved

1. John A. T. Robinson, *Time* (21 March 1977), 95.

2. John Warwick Montgomery, "Legal Reasoning and Christian Apologetics," *The Law Above the Law* (Chicago: Christian Legal Society, 1975), 88-89.

3. Louis R. Gottschalk, Understanding History, 2d ed. (New York: Knopf, 1969), 150.

4. Bruce, *The New Testament Documents*, 33.

5. Earnest Findlay Scott, *Kingdom and the Messiah* (Edinburgh: T and T Clark, 1911), 55.

6. *The Jewish Encyclopedia* (New York: Funk and Wagnalls, 1906), 8:508.

7. Joseph Free, Archaeology and Bible History (Wheaton, Ill: Scripture Press, 1969), 1.

8. E. M. Blaiklock, *The Acts of the Apostles* (Grand Rapids, Mich.: Eerdmans, 1959), 89.

9. Josh McDoweIl, *The Resurrection Factor* (San Bemardino, Calif.: Here's Life Publishers, 1981), 34-35.

10. Clark Pinnock, *Set Forth Your Case* (New Jersey: The Craig Press, 1968), 68.

11. Bruce, *The New Testament Documents*, 15.

12. Will Durant, Caesar and Christ, *The Story of Civilization* (New York: Simon and Schuster. 1944), 3:557.

Chapter 4 - What It Means

1. Josh McDowell, *Evidence That Demands a Verdict* (San Bernardino, Calif.: Here's Life Publishers, 1979), 328.

2. Paul Little, *Know Why You Believe* (Wheaton, III.: Scripture Press, 1967), 178.

3. Robert 0. Ferm, *The Psychology of Christian Conversion* (Westwood, N.J.: Fleming H. Revell, 1959), 225.

4. Nonnan L. Geisler and William E. Nix, *A General Introduction to the Bible* (Chicago: Moody Press, 1968), 24.

5. F. F. Bruce, *The Books and the Parchments*, rev. ed. (Westwood, N.J.: Fleming H. Revell, 1963), 88.

Made in the USA
Columbia, SC
26 May 2021